Studies in Modern History

General Editor: J.C.D. Clark, Joyce and Elizabeth Hall Distinguished Professor of British History, University of Kansas

Titles include:

James B. Bell
A WAR ON RELIGION
Dissenters, Anglicans and the American Revolution

James B. Bell
THE IMPERIAL ORIGINS OF THE KING'S CHURCH IN EARLY AMERICA, 1607–1783

Joe Bord
SCIENCE AND WHIG MANNERS
Science and Political Style in Britain, c. 1790–1850

Jonathan Clark and Howard Erskine-Hill (*editors*)
SAMUEL JOHNSON IN HISTORICAL CONTEXT

Edward Corp
THE JACOBITES AT URBINO
An Exiled Court in Transition

Eveline Cruickshanks and Howard Erskine-Hill
THE ATTERBURY PLOT

Diana Donald and Frank O'Gorman (*editors*)
ORDERING THE WORLD IN THE EIGHTEENTH CENTURY

Richard D. Floyd
CHURCH, CHAPEL AND PARTY
Religious Dissent and Political Modernization in Nineteenth-Century England

Richard R. Follett
EVANGELICALISM, PENAL THEORY AND THE POLITICS OF CRIMINAL LAW REFORM IN ENGLAND, 1808–30

Andrew Godley
JEWISH IMMIGRANT ENTREPRENEURSHIP IN NEW YORK AND LONDON, 1880–1914

William Anthony Hay
THE WHIG REVIVAL, 1808–1830

Mark Keay
WILLIAM WORDSWORTH'S GOLDEN AGE THEORIES DURING THE INDUSTRIAL REVOLUTION IN ENGLAND, 1750–1850

Kim Lawes
PATERNALISM AND POLITICS
The Revival of Paternalism in Early Nineteenth-Century Britain

Marisa Linton
THE POLITICS OF VIRTUE IN ENLIGHTENMENT FRANCE

Karin J. MacHardy
WAR, RELIGION AND COURT PATRONAGE IN HABSBURG AUSTRIA
The Social and Cultural Dimensions of Political Interaction, 1521–1622

James Mackintosh
VINDICIÆ GALLICÆ
Defence of the French Revolution: A Critical Edition

Robert J. Mayhew
LANDSCAPE, LITERATURE AND ENGLISH RELIGIOUS CULTURE, 1660–1800
Samuel Johnson and Languages of Natural Description

Jeremy Mitchell
THE ORGANIZATION OF OPINION
Open Voting in England, 1832–68

Marjorie Morgan
NATIONAL IDENTITIES AND TRAVEL IN VICTORIAN BRITAIN

James Muldoon
EMPIRE AND ORDER
The Concept of Empire, 800–1800

Julia Rudolph
WHIG POLITICAL THOUGHT AND THE GLORIOUS REVOLUTION
James Tyrrell and the Theory of Resistance

Lisa Steffen
TREASON AND NATIONAL IDENTITY
Defining a British State, 1608–1820

Lynne Taylor
BETWEEN RESISTANCE AND COLLABORATION
Popular Protest in Northern France, 1940–45

Anthony Waterman
POLITICAL ECONOMY AND CHRISTIAN THEOLOGY SINCE THE ENLIGHTENMENT
Essays in Intellectual History

Doron Zimmerman
THE JACOBITE MOVEMENT IN SCOTLAND AND IN EXILE, 1746–1759

Studies in Modern History
Series Standing Order ISBN 978–0–333–79328–2 (Hardback)
978–0–333–80346–2 (Paperback)
(*outside North America only*)

You can receive future titles in this series as they are published by placing a standing order. Please contact your bookseller or, in case of difficulty, write to us at the address below with your name and address, the title of the series and the ISBN quoted above.

Customer Services Department, Macmillan Distribution Ltd, Houndmills, Basingstoke, Hampshire RG21 6XS, England

---

Science and Whig Manners

Science and Political Style in Britain, c. 1790–1850

Joe Bord

palgrave
macmillan

First published 2009 by
PALGRAVE MACMILLAN

Palgrave Macmillan in the UK is an imprint of Macmillan Publishers Limited, registered in England, company number 785998, of Houndmills, Basingstoke, Hampshire RG21 6XS.

Palgrave Macmillan in the US is a division of St Martin's Press LLC, 175 Fifth Avenue, New York, NY 10010.

Palgrave Macmillan is the global academic imprint of the above companies and has companies and representatives throughout the world.

Palgrave® and Macmillan® are registered trademarks in the United States, the United Kingdom, Europe and other countries.

ISBN-13: 978–0–230–57484–7 hardback
ISBN-10: 0–230–57484–X hardback

This book is printed on paper suitable for recycling and made from fully managed and sustained forest sources. Logging, pulping and manufacturing processes are expected to conform to the environmental regulations of the country of origin.

A catalogue record for this book is available from the British Library.

Library of Congress Cataloging-in-Publication Data
Bord, Joe, 1977–
Science and Whig manners : science and political style in Britain, c. 1790–1850 / Joe Bord.
p. cm. — (Studies in modern history)
Includes bibliographical references and index.
ISBN 978–0–230–57484–7
1. Science and state—Great Britain—History—18th century.
2. Science and state—Great Britain—History—19th century.
3. Whig Party (Great Britain) I. Title.
Q127.G4B67 2009
509.41′09034—dc22 2008037191

10 9 8 7 6 5 4 3 2 1
18 17 16 15 14 13 12 11 10 09

Printed and bound in Great Britain by
CPI Antony Rowe, Chippenham and Eastbourne

Dedicated to my wife Nicole

Contents

Preface

This book brings a new optic to the study of Whiggery. Like any magnifying glass, the method of scrutiny expands the direct topic and tends to distort material around the edges. The reader is probably best advised to use other instruments first: the *Politics of Evolution* by Adrian Desmond; the *Age of Atonement* by Boyd Hilton, and *English Society, 1688–1832,* by Jonathan Clark are indispensable. I try to provide theoretical armature for the study of manners in political selection but the usefulness of the optic will really be judged by its application. Hopefully the reader will be convinced by the increased exposure of certain aspects of politics: the career of 'statesmanship', the importance of political generosity in the formation of liberalism, Whig agrarianism, anti-Catholicism and sociability.

Notwithstanding my responsibility for this book, I would like to record my full appreciation to Dr Boyd Hilton for his help and patience. I have also profited greatly from the influence of Dr Leslie Mitchell, and the intellectual stimulation and conversation of Dr Peter Mandler, Dr Simon Skinner, Dr James Secord, Dr William Ashworth, Dr Jonathan Parry, Dr Joanna Innes and Professor Frank O'Gorman. I would like to thank Dr Lawrence Goldman and Dr Simon Schaffer for their searching criticism and the anonymous reviewers of this book for their comments. I am grateful both to the Centre for the History of Science, Technology and Medicine at Manchester University for the hospitality and assistance it extended to me, and to Professor John Pickstone for facilitating this. I would like to acknowledge the support of the Arts and Humanities Research Council of the British Academy, and of Trinity College, Cambridge. Most of the research for this book was completed during my fellowship at Trinity (2002–06). I should also mention the assistance and access to resources given to me by Fordham University, Touro College, and the Nightingale-Bamford School in New York. Finally, the book is dedicated to my wife, Nicole Luna, without whom this work would never have been finished.

1
Manners, Science and Politics

Science and political manners

This book focuses on aspects of Whig politics in the first half of the nineteenth century. It considers some strands of science as they influenced Whig notions of how the statesman should appear. It is written not from the perspective of the history of science (the attempt to explain the pattern and development of scientific ideas and practices) but from the point of view of political history. In other words, the essay explores certain ways in which some manifestations of scientific engagement expressed the political identity of Whig statesmen. This kind of political history is not the same as the history of policy. Of course Whigs did sometimes bring science to bear on policy issues: an aspect treated elsewhere by the present author in relation to the Talents Ministry (1806–1807).[1] The circumstantial account attempted here has rather modest explanatory force when it comes to specific political actions, but it has value in indicating the context of politics. This is because it engages a note of thought and sentiment that was influential in political manners even while parliamentary choices were dominated by the great framing issues of public religion, social change, trade policy and parliamentary reform.

What does it mean to say that science was influential in political manners? Basically, that the use of scientific imagery in discourse – casual as well as keenly argued – was prominent in Whig circles, as was involvement in scientific bodies and modes of behaviour. At first sight, this might seem surprising. Outside of the Scottish intellectual universe, the 'Whig World' painted by Leslie Mitchell seems rather unscientific.[2] Yet even at the heart of whiggish society, there was lively interest in natural philosophy, as attested by Lord Holland's *Memoirs*.[3] Such subjects may not have been as central as constitutional history, or the classics, but they were present in ways that deserve further exploration. Science could be used to display values of generosity, sociability and cultivation. The contemporary notion of 'Science, which in its comprehensive sense means only *Knowledge*, and in its ordinary sense means *Knowledge reduced to a System*' was particularly congenial in this regard.[4] This old-fashioned *scientia* continued to resonate despite the disciplinary efforts that gathered pace during the 1830s. In 1828 Thomas Young complained to John Herschel that science was often used as little more than a nickname.[5] Whigs were producers as well as consumers of *scientia*. Such Whiggish usage was frequently dilettantish, but the pattern of selective use was characteristic and helped to demarcate Whig groups from other formations (for example, Ultra Tory or radical) that maintained their own manner of politics. The essay tries to provide some new perspectives on how this distinctive pattern worked. How to be a Whig mattered as well as the substance of Whig goals: science functioned not only as ideology or policy but also as style and display. In this, Whig science formed a sub-element of the broader genteel scientific culture. There was a great diversity of political objectives, alliances and ideologies among Whigs over the course of the first half of the nineteenth century. There was also a recognisable cohesion of political manners, and science played a discernible part in this continuity.

The claim of an underlying and long-lasting similarity of Whig manners c.1790–1850 needs to be justified. After all, there were generational, intellectual and temperamental differences between groups such as Charles James Fox's metropolitan circle, country Whigs such as Coke of Norfolk and the morally serious Young Whigs of the next phase identified by E.A. Wasson.[6] The Whig–Liberal Reform platform under Russell's leadership was much larger and more diverse than Fox's opposition fraction, even though those returned within the party as 'Whigs' in 1847 were comparable in number.[7] Half a century is a long span, and we would expect variation over time. Yet the fact remains that during all of this period groups claimed the Whig mantle and so it is reasonable to suppose that there were continuing factors. So far as continuity is concerned, what matters is not whether Foxites in 1802 would have hypothetically recognised 1847 Whigs as brethren, but whether 1847 Whigs did in fact claim Foxite descent and acted in ways that recollected their political forerunners. The argument that follows discusses some similarities of manners that essentially operated as means of political recognition. These are grouped under four broad headings corresponding to subsequent chapters of the book. The manners discussed had substantial continuity over a long period and lay in parallel to religious, ideological and practical differences among Whigs. This does not mean that the similarities were more important than the differences in determining specific political choices. It was perfectly possible for politicians to recognise fellow-Whigs (or claim descent) and then to disagree on many points of policy or doctrine. Moreover, despite overlap and competition, the overall pattern was not shared by radicals and Tories, even by Liberal Tories and their Peelite successors. Indeed, the effective amalgamation of the greater part of the Peelites with the Whigs in the early 1850s coincided with the dilution of some, if not all, of the Whig manners under discussion. The argument implies that the realignment of the 1850s might have been more important in affecting Whiggish manners than the Whig assumption of power in the 1830s. The governments of Grey and Melbourne were culturally in opposition (to the chagrin of the latter minister) even when they were in power.[8]

The manners explored here should be understood as ways of identifying attractive values in public behaviours. The first of these was liberality, in the sense of being generous and munificent. The second was the advantage in showing one's character as a statesman, a public man of comprehensive knowledge. The third was the need to cultivate and be cultivated, as influenced by a georgic tradition; and the fourth was the imperative to display a particular sort of rational sociability. Of course these mannered values did not exhaust Whig mores. However, science was recruited to express these values in particularly relevant ways. Moreover, they remained relatively stable while other Whiggish values (such as progress, improvement and toleration) were subject to greater change precisely because they were more immediately bound up in policy disputes. These other values were more significant ideologically. Nevertheless, so far as the formation of a public Whig persona is concerned, it is helpful to focus on some values which have been occluded by headline ideology, namely liberality, statesmanship, cultivation and rational sociability.

If ideology is the system of values that mobilises political action, manners (or style) comprise those values whose expression conveys group identity. Questions of signalling, positioning and the display of public personae become central. Manners are defined here as public shared behaviour current in political society. They form part of the background of politics, and their study generates topics of fashion and grouping which are ill-served by chronological approaches to issues and ideas. Manners are ways of consolidating groups and communicating networks through shared speech and behaviours. They are learnt, and often involve conscious anachronism and the echoing of a selective past. Therefore they are suited to a thematic rather than a narrative discussion. As manners existed to convey an impression, the appropriate treatment is to take a number of suggestive examples illustrating different facets of display. The episodes chosen here might seem miscellaneous, but they are selected to expose cases of style where science was evoked to express political identity. Hence there is a lot more here on Brougham than on Grey, Russell, Althorp and Landowne rather than Fox. Not all sciences are well represented, and the cases punctuate the period rather than forming a single continuous story.

What are these cases, and why are they chosen, given that by no means every Whig interaction with science is treated? This book is really interested in underestimated aspects of Whig political identity, rather than the sciences themselves, or even the Whig use of science *per se* across the range of ideologies and policies. It spends time discussing the character of styles such as statesmanship, liberality and the georgic tradition and then notices certain manifestations of scientific engagement that illustrate these values at work. Politicians come into the account as they seem to typify this or that manner. Thus Lansdowne, Holland, Althorp, Brougham, Russell and the Bedford connection do not provide a complete gallery of Whig leadership; instead they help to show some manners among the Whigs. These include the pose of being in political life but somehow transcending it; of resisting arbitrariness through scientific republicanism and rejecting hypothesis; of expressing a political ideal of party through the rational coalition of a scientific society; of resisting incoherence by opposing irrational (and Catholic) bodies and of cultivating improvement through fertility. The exercise is therefore rather peculiar and does not fit easily into any one historiographical category. It probably comes closest to a history of ideas, but with the uncomfortable awareness that its themes comprise more than ideas or ideology. The argument does not set out to prove a negative or pretend to be exhaustive and it shows no more than its examples indicate.

All of this does prompt a basic methodological challenge: Is such a study properly history at all? Is it not instead a form of anthropology? The approach goes back to Clifford Geertz's definition of culture as structures through which participants can do certain social things like conveying insults or concluding ententes.[9] We want to try to expose intellectual conventions and etiquettes. Undoubtedly, if one had access to the kinds of intensive, interactive study offered by ethnographic methods one would deploy them, but the period and the problems of evidence entail historical investigation. One should admit from the start that one of the difficulties besetting the present exercise is that the phenomena highlighted do not always leave direct source evidence – manners are best seen, experienced and reported directly. To put the problem another way, there is a discrepancy between the kind of phenomena we want to catch and the available methods. If most of the evidence is by necessity textual and discursive in nature, it does not mean that the subjects under discussion were restricted to text or discourse. The situation is eased somewhat by the fact that publications and especially letters were themselves forms of mannered behaviour. Manners may not have been as important as ideology or class in driving events, but they existed and are worth trying to understand, even though they are often elusive. The Whigs are particularly interesting as an example of political manners because their identity was a matter of culture to a degree not evident in current political formations. Being a Whig was a way of having a politically defined life for an influential subset of the British elite for a protracted period.

Whiggery and the political narrative: An overview

Being Whig was also the default political identity of the dominant spectrum of politicians from the accession of George I up until the 1790s. That decade began the decisive breakdown of Whiggery as the dominant political identity just as the 1760s had witnessed the practical disruption of the Newcastle Whig connection and its allies. Most of the rival groups in the first part of the reign of George III regarded themselves as Whigs of one sort or another. Pittites, Northites, Portlandites, Foxites and Grenvillites all claimed to be Whigs. The conflicts of the French Revolution began a process of contraction so that by the 1820s only the Bedford–Foxite opposition led by Charles Grey constituted the 'Whigs'. Conversely, the Pittites and conservative Whig factions that had joined the government largely over the war (Portlandites, most Grenvillites) came to be thought of as 'Tories'. These new groupings under old nomenclature were both affected by novel intellectual and ideological developments, of which Romanticism and especially Liberalism were the most important. Historians have made progress in appreciating the complexity of 'Liberal' currents, which will have to be considered in due course, but at present it is enough to note that their spread put particular strain on the governing Tory constellation, creating the possibility of liberal–centrist realignment. Canning's death in 1827 destroyed one such opportunity but Grey's coalition of 1830 brought the leading Liberal Tories Palmerston, Grant, Ripon and Melbourne into a Whig government. There were tensions between the Foxite Whig and liberal portions of the Reform administrations, particularly over social policy, but these were not so great as the contradictions between liberals and Protectionists (including most anti-Catholics) inside Peel's Conservative party, which returned to power in 1841. These contradictions burst apart the Conservative synthesis during the prolonged crisis of 1845–1846 (Maynooth followed by Corn Law Repeal). After this the preponderance of the Peelite wing eventually amalgamated with the Whig-Liberal party, which thereafter dominated politics for the next twenty years.

Naturally enough, political historians have been concerned to explain the course of this political narrative. Up until the 1970s their focus was on high political factors: party manoeuvre, the pursuit of tactical advantage and ambition for place. Divisions of principle were strongly tied to concrete issues such as parliamentary reform and (down to 1829) Catholic Emancipation. Parliamentary reform suggested a spectrum that placed Whigs and Liberals between Radicals and Tories/Conservatives. The most thorough exponent of this approach was Norman Gash.[10] However, over the past thirty years or so, there has been an intellectual-historical revolution in the historiography incorporating a stronger emphasis on religion and social thought. Thus Peter Mandler has identified a distinctive Whig social ideology competing for predominance with classical liberalism in the Reform politics of the 1830s–1840s.[11] Struggles over, for example, Poor Law Reform (1834) move to centre stage with this interpretation. Richard Brent has investigated the Liberal Anglicanism of a generation of Whigs who succeeded Charles Fox's immediate contemporaries.[12] Liberal Anglican aspirations towards comprehensive and tolerant erastianism provoked strong reactions in areas such as tithe reform and education. Brent is particularly successful in bringing into play an intellectual current that had previously been designated in largely literary terms.[13] John Burrow has gone on to distinguish Whig hierarchical and pluralist social philosophy from liberal individualism, a distinction evident in the politics discussed by Mandler.[14]

Burrow's approach to the political history of ideas is comparable to J.G.A. Pocock's method of analysis.[15] Both of these historians engage in the dynamic interpretation of clusters of text, leading to sophisticated discursive histories. This has purchase for political history, because from Edmund Burke onwards many influential political figures were also authors (Macaulay, Russell, even Holland on the Whig side; Gladstone and Disraeli among opponents). For Pocock, the intellectual history of Whiggism is part of a broader sweep of Atlantic political thought and is relevant to the long development and eventual dissipation of a civic republican or humanist tradition.[16] Both Burrow and Pocock identify the significance of the transition in meanings between the commercial society of the eighteenth century, with its culture of politeness, and the bourgeois, progressive and industrial articulation of nineteenth-century society. Finally, another important contribution to the intellectual-historical turn in political historiography must be mentioned at this early stage. Taking his cue from the conflicts between liberals and high tories, Boyd Hilton has discerned a clash between organic and mechanical philosophies driven by the uncertainties unleashed by the French Revolution.[17] The playing out of these philosophies was heavily influenced by permutations of evangelicalism and the religious revival under way since the 1780s. These were all-encompassing responses to a period of turmoil, making sense of politics, human behaviour, the natural world and the divine nature. Science had a clearly ideological part to play in this polarisation, acting as a crucial field of debate between opposing homologies.

Now, this book does not make a direct contribution to explaining the political narrative. It furnishes an *indirect* supplement to the more recent historiography in trying to address a rather different political question: the problem of manners and political recognition. How did different groups know who was on their side? How did individuals demonstrate their allegiance to various connections and networks? What type of persona did one have to cultivate in order to be a Whig? The book focuses on the contribution of science from the angle of its presence in manners. Ideology and discourse come into this as mannered behaviours. Granted, it is neither necessary nor desirable to be too rigid in policing the boundary between ideology and style, because values could have all sorts of overlapping functions. The perspective certainly does not preclude or replace the properly ideological role of science as an intellectualisation of political concurrence and conflict. However, it does round out our view of the early-nineteenth-century political world by indicating that science did not just function as ideology. This approach has been stimulated by Leslie Mitchell's studies of the Foxites and Holland House. Mitchell concentrates on Foxite sociability and an almost ritualised reliance on obsolete political categories. His Whigs re-fight the battles of 1784 in 1834. What Mitchell's approach really exposes is a second dimension of Whig politics: the cultivation of politico-cultural identities as sufficient ends. The implication is that Whig politicisation was not exhausted by the pursuit of Whig political objectives. The criticism of Mitchell here is that his range of manners is too narrow. As well as racing, gambling, dining and literature, one needs to pull in other activities including, for the purposes of this book, science and cultivation.

To some extent a similar expansion of view has been performed in relation to political economy by Peter Mandler in his *Aristocratic Politics*. The styles identified by Mandler are oriented towards policy differences, as indeed are the intellectual homologies explored by Hilton. Divergent ideologies demarcate factions with different views on the resumption of cash payments, free trade in corn, Poor Law Reform, education policy and the rest. Yet not all styles, or manners, were closely linked to policy choices. Rather they helped to reinforce the groups that then constituted the subjects of politics. The reorientation of groups could have long-term consequences. One of the peculiarities of the Whig story is the way in which the accession of most of the leading Peelites in the years after 1846 actually accelerated the decline of Whiggery within British Liberalism. The augmentation of Whig–Liberal forces did not ultimately lead to the consolidation of Whig power. The process was prolonged by Lord John Russell's premiership (1846–1852), but by the early 1850s it was clear that Whiggery was going out of fashion. Russell's literary rearguard action – the issue of his *Memorials and Letters of Fox* (1853) and *Life and Times of Fox* (1859) – recapitulated a cultural history that stretched back to the 1790s but had decreasing resonance in a party dominated by Viscount Palmerston (1784–1865) and William Gladstone (1809–1898). Even though the Whig social conscience did not have the monopoly of legislation in the Reform administrations of the 1830s–1840s, Mandler argues quite convincingly that Whig government (typified by Russell's encouragement of the 1847 Factory Act) was the moment of an aristocratic public ethos that did not effectively outlive its apotheosis. It is true that T.A. Jenkins has pointed to the continued presence of Whig–Liberals in the 1870s and 1880s, chief among them being those keen natural historians, the Marquess of Hartington and Duke of Argyll.[18] However, the break-up of Lord John Russell's intermitted administration in 1852 constituted a definite punctuation mark. Russell was indisputably perceived as a Whig: the Last Doge. Gladstone certainly was not. There was a crucial loss of cultural capital and prestige among Whig grandees right at the point when new forces were seeking to redefine British Liberalism. Jonathan Parry has argued persuasively both for the success of the Whig-led Liberal coalition and the ultimate failure of Whigs to protect their synthesis initially from Gladstone, and later from Joseph Chamberlain.[19] This process of going out of fashion was anticipated by the decline of Whig manners of science, liberality and cultivation.

The variety of Whiggery: Further aspects

Whig politics rarely signified propositional coherence, and the imprecision of the word 'Whig' in this sense has led Parry to describe the lack of consensus over who counted, referring the term after 1850 to Liberal party traditionalism or a genteel social position.[20] Parliamentary historiography has tended to define the kaleidoscope of Whiggism through the narrative of its successive shifts, especially in retrospect, through the story of an ending or merging into wider Liberalism (and ultimately out of Liberalism into Unionism). Writing in a broadly Namierite register, political historians such as Gash and Leslie Mitchell have described how Whig politics before the mid-nineteenth century was composed of a series of contingent family alliances and friendship circles.[21] This kind of analysis has a lasting value, helping one to recognise that dynastic topography framed intellectual commitments: not only, erratically, to 'peace', 'retrenchment' and 'reform', but also to liberal styles that absorbed science.

Thomas Macaulay proclaimed that the science of government was both experimental and progressive.[22] We need to consider how such arguments, as well as being ideological statements, acted as positioning signals in the political firmament. There was a rough correlation between increasing political attention to natural science and the rise of political Liberalism. On the one hand, there was a correlation between the rising political profile of science (with the involvement of many Whigs) and the emergence of Liberal politics. On the other hand, this incidence was not confined to Whig connections; neither were all prominent Whigs so engaged, with Charles Fox and Charles Grey, for example, showing little appetite. The esoteric speculations of Tory MP and squire Sir Richard Vyvyan, and weighty patronage provided by Sir Robert Peel (1788–1850), point to other strands in a much wider intellectual rapprochement.[23]

What was the background to this developing economy of signals? Science had never been a Whig monopoly, and during the previous century its political importance fluctuated. A rich tradition of estate-based experimentation had strong continuity throughout the eighteenth century. There are tantalising glimpses of the second Marquess of Rockingham's experiments on coal tar and, more obviously, the ubiquitous gentlemanly interest in crop productivity, livestock and pasturage.[24] Yet Isaac Kramnick argues that the English political and constitutional legacy was the most significant factor in determining the contours of earlier Hanoverian debate.[25] Court Whigs of the first rank such as Sir Robert Walpole and the Pelhams, (and also Pitt the Elder) were not intensively engaged in natural philosophy.[26] The vital centre of political argument lay elsewhere. One could not say of the rising political generation of the 1730s, as Brougham did of his Edinburgh associates in the 1790s, that they were obsessed with chemistry and political economy.[27] Even the invocation of Newtonian imagery by politicians was more belated than one might have expected; the heyday of constitutional mechanics came with Blackstone's *Commentaries on the Laws of England* (from 1765).[28] Newtonianism pervaded the polite and enlightened culture of mid-eighteenth-century Britain, but perhaps science was less conspicuous in party and parliamentary politics than in the subsequent period. Benjamin Franklin wrote in 1769 that the Royal Society comprised all parties and yet managed to be of none.[29] The 'Club of Honest Whigs' (in the 1760s and 1770s), which combined convivial discussions of philosophy and politics, included leading dissenting teachers and clergymen and luminaries such as Franklin, Joseph Priestley (1733–1804) and Richard Price (1723–1791), but no parliamentarians.[30]

On the other hand, the more one looks at the Augustan age, the more science in politics one finds. Leading Whig parliamentarians of the Walpolean and Newcastle era were not indifferent to the patronage and ideological possibilities of natural philosophy. D.P. Miller has pointed to the existence of a Hardwicke circle at the Royal Society between the eras of Newton and Banks.[31] This puts Franklin's remark into context. The first and second Earls of Hardwicke were firm Newcastle Whigs and the second Earl, Philip Yorke (1720–1790), carried this over to support of Rockingham. Their connection generally prevailed in controlling the senior positions at the Royal Society until the advent of Joseph Banks in 1778. Before the 1770s it was sometimes possible to articulate party divisions through natural-philosophic ideologies. Miller notes how Hardwickian patronees were primed to attack non-juring Hutchinsonianism and were involved in controversies over the public electrical demonstrations identified by Simon Schaffer.[32] There were rival Whig and Tory enthusiasms for Newtonian models.[33] Certain opposition republican theorists flirted with visions of cosmic disorder, and Bolingbroke's links with the French Newtonian school may have generated some equivalent imagery in the *Dissertation upon Parties* (1733–1734).[34]

Larry Stewart and Margaret Jacob point out that Newton's own patronage at the Royal Society from 1703 was underpinned by his Whig connections, which gained him preferment, particularly at the

Mint.[35] He was given a state funeral at Westminster Abbey in 1727 by the Walpolean regime. Newton's popularisers among the Boyle lecturers were Court Whig and latitudinarian theoreticians such as Bentley, Derham, Whiston and (most importantly) Samuel Clarke. The Newtonian lecturers were an ideological prop to the Hanoverian side during the 1715 crisis over the succession of George I.[36] Stewart has explored a number of ways in which Whigs patronised science in the early eighteenth century. The Whig Junto under Queen Anne assisted lecturers William Whiston and Francis Hauksbee in 1710–1713. Stanhope and Walpole defended Whiston's longitude proposals, culminating in Whig promotion of the Longitude Act (1714).[37] Tory vituperation in the hands of Sacheverell and Swift condemned whiggish modern philosophy.[38] One significant Whig patron who anticipated later magnate involvement was James Brydges, the Duke of Chandos, who cultivated a network of technical entrepreneurialism.[39] However, Stewart draws back from identifying Newtonianism with Whiggery too closely, pointing out counter-examples of tory engagement.[40] He argues that the practical demands of experimental philosophy, of market and self-interest, in the end overpowered ideological factors.[41] It is striking that although his period extends to 1750, most of Stewart's examples of Whig patronage are drawn from the polarised years before 1730. The ideological conscription of science began to decline after the entrenchment of the Robinocracy.[42]

Natural philosophy regained a sharper political edge during the period of the American Revolution and the contemporary wave of domestic reform movements. The constitutional implications of Unitarianism became clearer from the Feathers Tavern Petition (1772) onwards. Priestley supported the movement against compulsory subscription to the Thirty-Nine Articles and was a close ally of petition organiser Theophilus Lindsey (1705–1787). Historians of dissent have shown how sharp differences of philosophical doctrine underlay the emerging cluster of reform ideas. Priestley's materialism and necessitarianism contrasted with the moral Newtonianism of James Burgh and Price, derived (as Priestley's thought had been originally) from Samuel Clarke's earlier analogical natural theology.[43] Priestley and Price were significant in the circle of Lord Shelburne, but not in the circle of his son, the third Marquess, explored below. This aspect of the English Enlightenment turned out to be less influential on the theory and practice of the Whig party than the Scottish philosophical tradition (which in Brougham's hands expressed itself as a militant Newtonianism) or a Whiggish variant of agrarian patriotism.

A development in the use of science to signal political identity occurred in the 1770s with the enlightened patronage of the Earl of Shelburne (1737–1805), the sponsor of Jeremy Bentham and Joseph Priestley, and father of the third Marquess of Lansdowne, as described by John Norris.[44] Yet Shelburne's circle was most unusual, and Shelburne himself cut an isolated figure among his political peers. It was only during the 1790s that the Foxite opposition attracted a high concentration of philosopher-politicians, including Shelburne (by now the first Marquess of Lansdowne), and the 'Citizen Lords' Stanhope (1753–1816) and James Maitland, eighth Earl of Lauderdale (1759–1839). The last of these was significant for his sponsorship of the rising generation of Scottish, predominantly Edinburgh Whigs: men such as Francis Horner (1778–1816) and Henry Brougham (1778–1868), who went on to contribute to the emerging reform politics. Just as important was the active experimentation and speculation of the English aristocracy and gentry: of Viscount Althorp (1782–1845), for example, and the Bedfords, the latter patrons commonly reduced in secondary accounts to the passive dispensers of largesse. Another neglected group were ministerial Whigs who uneasily straddled opposition and government at different times. William Wyndham, Lord Grenville (1759–1834), is particularly interesting as an example of this type. Below the commanding heights were reformers including the Liverpool Foxite-botanist William Roscoe (1753–1831) and the radical Whig Henry Grey Bennet (1777–1836), member for Shrewsbury and enthusiastic gentleman geologist.

The gallery of statesmen can, of course, be classified in different ways. Henry Petty-Fitzmaurice, the third Marquess of Lansdowne (1780–1863), was one of a generational cohort whose correspondence and literary remains occupy a large slice of this study. This cohort includes Horner, Brougham and Althorp as well as Lansdowne. Lord John Russell (1792–1878) can be counted as a cadet member of this generation, since he combined a Scottish education with a Foxite lineage. All except Lansdowne have had their political biographers; but none of the latter has got to grips with the political resonance of their natural science. This is largely to do with the narrow focus of most parliamentary historians, but is also related to the limitations of the biographic form. The most prolific writer and experimenter of the four, Henry Brougham, is particularly ill served, with one biographer, Robert Stewart, dismissing his science as a form of relaxation and diversion.[45]

The Scottish, and particularly the Edinburgh, Enlightenment is too often treated as a discrete entity, with lines of influence emanating southwards from the Athens of the North. Whigs were and are often condemned for their social condescension, and yet their concept of statesmanship was at root an egalitarian one that could be made to accommodate persons of relatively humble social status. Statesmanship had a universal aspect. It implied a kind of social mobility through comprehensive knowledge, patriotism and cultivation of the laws. For the ambitious generation of Edinburgh alumni at the turn of the century, the challenge was to locate their political identity by using particular means of cognitive display. To put it another way, because the persona of the statesman increasingly came to be defined by its cognitive content, the carriers of the Scottish enlightenment legacy could lay claim to a distinctive set of intellectual resources. In November 1802, Francis Horner noted in his journal that

> Political Science can only be prosecuted after the manner of the Baconian logic, and the practice of that must be most effectually learned from good models; it will be of inestimable value indeed, if to those we already possess in some of the physical sciences may be added existing models of philosophic investigation relative to mind.[46]

This is a clear statement of methodological intent, and can be explored as such. Yet in the same journal entry, Horner copied down two passages from Stewart's *Memoirs* that 'awakened a train of personal reflections': '*the happiness of a liberal occupation, superior to the more aspiring aims of a servile ambition*'; 'but I too, have designs and enterprises of my own . . . '[47] Horner resolved to pursue a 'mixed' line of activity. Philosophic politics involved tensions for its practitioners. Scottish theory was undoubtedly significant, but it is possible to overstate its internal sufficiency over the pull of wider ambitions.

The Scottish legacy was also continuously amended by distinct administrative and patrician experiences. Lansdowne (then Lord Henry Petty), Horner and Brougham were university contemporaries at Edinburgh in the later 1790s. Petty and Althorp were both up at Cambridge in the very first years of the new century, the former beating the latter to Pitt's old Cambridge University seat in 1806. Apart from Horner, who died prematurely, the members of this group held high office in the Reform government of November 1830. Lansdowne manifested a technical, administrative form of liberal politics. His Lord Presidency of the Council was matched by Brougham on the Woolsack, and by Althorp's central tenure of the Chancellorship of the Exchequer and Leadership of the Commons.

Meanwhile, the context and framework of Whig politics continued to be set by insular history. The cohesion of core clan loyalties – of Spencers, Cavendishes, Russells and relative latecomers like the Hollands – corresponded to distrust of the crown.[48] This suspicion had originated in George III's disruption of the Hanoverian Court Whig system in the 1760s, and lingered on into the Reform ministries of the 1830s.[49] It was especially enshrined by memories of the constitutional crisis of 1782–1784, which had propelled Pitt to power. If Pitt occasionally called himself an independent Whig, it was to express his alienation from many of these networks, which were descended from the Newcastle and Rockingham connections of the mid-eighteenth century. Shifting alliances might be struck depending on political exigencies to check the throne. Thus the Lansdowne interest became a permanent fixture of Whig politics from 1793 onwards as Shelburne sunk his differences with Fox to oppose the French Revolutionary War. This alignment solidified under the aegis of Shelburne's second son, the third Marquess of Lansdowne, in 1809. However, many Rockinghamites broke with Fox between 1792 and 1794 to support or take wartime office under Pitt, including the Marquess of Rockingham's heir and nephew, the Earl Fitzwilliam (1748–1833) and the Duke of Portland (1738–1809). The drift back towards the Foxites was driven principally by the need to secure Catholic Emancipation, a measure that Pitt had tried and failed to obtain against royal obstruction in the wake of Irish Union in 1800–1801. Grenvillites who had worked with Pitt since the 1780s allied with Fox on this basis, in 1802, and maintained an uneasy axis until Lord Grenville lent his support to the suspension of habeas corpus in 1817. Pro-Catholic Emancipation Pittites such as George Canning (1770–1827) decisively rejected the Foxite cousinhood. Nevertheless, Canning was eventually prepared to concert with the moderate Lansdowne wing of the Whig party in 1827 in order to forestall the Ultra anti-Catholics. The alignment was momentary, and eclipsed by Canning's death, but it cleared the way for close cooperation between Lansdowne Whigs and Huskissonian liberal tories.[50] This was followed by the accession of Canning's former-Foxite Irish Secretary William Lamb, later by Viscount Melbourne (1829) and most of the other leading Canningites (for example, Grant and Palmerston) to the governing Whig alliances of the 1830s.

Aristocratic Whig politics was thereby constituted by a range of formations, each with its own priorities and discursive preferences. The Pittite and Liberal Tory caricature of Foxite amateurishness and ignorance of political economy did not apply to the Lansdownes or the

Grenvilles, for whom administrative liberalism and effective executive power were central concerns. This did not prevent Grenville and Lansdowne from clashing directly over the coercion of domestic dissent in 1819. Grenvilles and Foxites cooperated in the cause of Catholic Emancipation, but the resumption of war with Napoleonic France after the Peace of Amiens (1802–1803) caused tensions as did Grenvillite aversion to parliamentary reform. Although the Foxite leader Charles Grey (1764–1845) became tepid about his earlier resolve to guide and contain pressure for widening the franchise and redistributing seats in the House of Commons, parliamentary reform remained unfinished business. This, and a new pious preoccupation with free trade and the corruption of the public realm, became the hallmarks of a younger generation of noble politicians in the 1820s: scions of the core Whig dynasties such as Lord Althorp (1782–1845), Lord Tavistock (1788–1861) and Lord Milton (1786–1857).[51]

The reach of Whig politics radiated beyond the aristocracy. From the 1830s onwards Whigs were forced to negotiate with Radical and Irish formations and, after 1846, with dissident Peelites. Between the Fox–Portland split in 1794 and Canning's coalition in 1827, different Whig aristocratic networks sought to lead, and intermittently to patronise, three broad groups of non-aristocratic opinion.[52] The first were predominantly urban, popular Whigs, such as those of the 'Mountain' headed by the brewer Samuel Whitbread (1764–1815). This group faded after its leader's suicide, but revived more informally in the 1820s. It lobbied conspicuously for anti-slavery, pacifism, reform of parliamentary inequities and ministerial abuses and an overhaul of the criminal law and the prisons. On many of these issues they shaded into Burdettite or Benthamite radicalism. The second was a small though prominent set of Scottish intellectuals and professional men, including Horner and Brougham. These Scottish 'philosophic' Whigs were often in the forefront of questions of commerce and political economy, illustrated by Horner's currency crusade culminating in the Bullion Committee and its Report (1810–1811). Philosophic whiggery has been heavily worked over by historians, its intellectual importance outweighing its small numbers of active politicians.[53] Most numerous of all was the body of provincial gentlemen and country politicians, typified by the Norfolk worthy, Thomas Coke of Holkham (1752–1842). Leslie Mitchell correctly identifies the disdain for country pursuits in Holland House, but his account neglects the agrarian side of Whig politics.[54] A bias towards cheap government and traditional liberties, coupled with wariness about free trade, fund holding and comprehensive parliamentary reform, characterised this backbench stratum. The leadership's efforts to mobilise these country Whigs reached a high point with the defeat of the property tax in 1816, and great importance was attached to their support throughout. It was an underlying reason why Whig magnates were so keen to foster shared interests in agricultural improvement and cultivation.

The symbolic and practical importance of the country strand has been skated over in conventional accounts of Whig politics. Yet it was acknowledged even by those usually considered as embodying a very different kind of politics in the historiography of the Whig party. The point is reinforced by an example: the parliamentary diary of Henry Grey Bennet, prominent member of the popular-Foxite Mountain and President of the London Geological Society (1813–1815).[55] The extant volume of Bennet's parliamentary diary begins in January 1821. This source has been neglected in histories of the politics of the Mountaineers, perhaps because it contradicts the picture of an urban and mainly metropolitan liberal radicalism.[56] While it is the case that the organisation of the Mountain disintegrated during 1815–1817 in the aftermath of Whitbread's death, many of the participants, including Grey Bennet, remained active and loosely cooperated with one another. Given Bennet's Rockinghamite emphasis on fighting the Crown and his musings on the consolidation of monarchical power over the crises of fifty years, it is better to see him among the radical wing of the Whig party, rather than as a protagonist of a separate Radical bloc. His evidence is a corrective to William Thomas' attempt to translate the politics of the Westminster constituency (with its radical-versus-Foxite polarisation) to the national scene. A telltale sign of Bennet's Rockinghamite-Foxite (rather than popular-radical) provenance is his dating of the egregiousness of Crown toryism to the accession of George III. It is also worth noting that there is no trace of a Norman Yoke or anti-Tudor argument in his political journal.

Bennet drew a fundamental distinction between false country gentlemen (thrown up by the government to deflect committee criticism), selfish Tory squires (who were concerned with their own rents and who could ultimately be bribed by the government) and liberal gentlemen such as his friend Charles Callis Western (1767–1844).[57] It was right to safeguard direct food supplies and to relieve the landed interest, but not at the cost of profligate spending and high taxation, nor should prices be driven upwards by agricultural protection (though Western had in fact advocated a form of variable protection in the Corn Law of 1815).[58] The proper course of reform for truly liberal gentlemen was economy in government and the reduction of taxes. These were standard positions; what is worth noticing is Bennet's insistence that country gentlemen could be liberal gentlemen also. The diary describes Bennet's frustration in 1821 as he discovered that most of the gentlemen were indeed selfish or sham, and biddable through particular concessions (such as repeal of the horse tax) rather than being willing to be marshalled for retrenchment.[59]

The Whigs were less effective horse-traders than their political rivals. Pragmatic administrative technique was also at variance with the sensibility of English Rockinghamite whiggery. Both aspects of practical politics were overshadowed by history. The Whiggism originally captured by Fox was a politics of commemoration and sacred myth. Leslie Mitchell observes that while the Pitt–Portland alliance represented a new alignment of property against republicanism, for the Foxite party what mattered was the perennial struggle with the executive as corrupted by George III.[60] In turn, the imagined genealogy of this contest was traced back to the cause of Hampden and Sydney.[61] Austin Mitchell's generalisation that Whigs embraced historical processes and eschewed abstract theories was certainly true of the Foxite heart of the party.[62] The significance of Foxite memory was grasped by its radical critics. The taunts of 'No Coalition' that greeted Fox at the Westminster election of 1802 were wounding precisely because Foxite cooperation with the Grenvilles recalled the unlikely cohabitation with Lord North in April–December 1783.[63] In 1812, William Cobbett enquired whether the Foxite principles that the Whigs were accusing the Prince of Wales of abandoning were principles held by Fox in office or contrary ones in opposition.[64] As late as 1827, Francis Place was to write to John Cam Hobhouse to complain of the inadequacy of a politics based on memory: invocation of 'Wilkes and Liberty' could not compare with the test of utility.[65] The implication was that Foxites did not understand legislation adjusted to this latter sober criterion.

The histories forged under political pressure by Lord John Russell, Lord Holland and Charles James Fox were distinct from the Whig History produced later by writers such as Macaulay and Trevelyan. It is tempting but misconceived to concentrate upon the unbroken Whig conversation that existed from Hume to Acton. Of course it is possible to trace such a genealogy, but only at the cost of passing over the specific pressures of parliamentary and social conflict on practising politicians. John Burrow concedes that such an evocation of a continuing political vocabulary brings with it the penalty of forgoing certain particular contexts.[66] What was at stake in the fifty years after the French

Revolution was politically effective memory. Richard Brent has captured, for instance, the way in which Russell deployed the historical narrative to moralise Foxite politics in the post-Napoleonic world.[67] The Whig flirtation with natural philosophy bore strong resemblances to this opportunistic manipulation of historical memory. Like their political histories, the scientific styles adopted by Whig statesmen evinced certain pronounced biases, and were not entirely incoherent, but they manifested a grabbing of cultural material rather more than the careful working out of methodological or philosophic positions.

John Pocock has pointed to the existence of a strain of ancient constitutionalism that was periodically revived by the anti-war Foxites (1794–1815) and only incompletely absorbed by later progressive syntheses.[68] For such Old or True whiggery, the present was adjusted to a virtuous past, which functioned as a regulative ideal. Other whiggisms had distinct conceptions of what progress might entail. During the 1820s, for example, Ricardian liberalisation was advocated by John Ramsay McCulloch (1789–1864) in the Whig-leaning *Edinburgh Review*. McCulloch subsequently advised Sir Robert Peel in the 1840s.[69] Meanwhile, the Foxite tradition, carried forwards by Lord Holland (1773–1840) and Charles Grey focused on the balance of the constitution, a preservative tension that prevented the easy falling away into absolute government.[70] There was a powerful declinist current in this strand of Whiggism, a sense of the steep constitutional odds against which improvement had to be won.[71] Reform may not have been conceded to popular pressures, but it was marked by intellectual defensiveness against crown corruption and, therefore, perennial pessimism. In Richard Brent's account of the religious rejuvenation of Whig politics, a younger generation of Liberal Anglicans broke through the dead end of Foxite mechanical constitutionalism, superseding theologically empty notions of a balanced polity in pursuit of the popular moral good.[72] Brent perceives a fourfold division among whig-liberal ranks, ranging from secular statesmen mainly comprising the older Foxites, through to Liberal Anglicans such as Lord John Russell, evangelicals and high churchmen. By comparison, Mandler stresses the social and economic distinction between interventionist Whigs and those liberals who were firmer believers in the beneficence of unimpeded exchange.[73] To complicate matters further, Hilton has pointed out the influence of moderate evangelicalism upon laissez-faire politics. This defined the Liberal Toryism of the 1820s, but was sometimes present on the Whig side also.[74] Such influences weighed against historical progressivism, emphasising instead the static and retributive justice of social transactions. From the perspective of political historiography, the positioning of Whiggery in terms of conservatism and forward-looking radical reform is therefore at best a partial understanding, if a persistent one.

Whiggery was not simply what Whig politicians did; it was the product of a constant process of self-examination and the conscious declaration of political identity. The peculiarity of Whig reflexivity deserves prominence. It is the counterpoint of J.C.D Clark's acute observation that the three ministries that dominated the period between 1770 and 1827 – those of North, Pitt the Younger and Liverpool – were faced with Whigs acting as an oppositional party without themselves possessing a central party character.[75] It was not until Melbourne's rejection of coalition with Peel that a two-party system crystallised (holding until the Corn Law crisis of 1846). Well into the 1830s the basic Whig register was one that linked party to opposition rather than to government. The need for self-definition outside the orbit of power was exemplified in Brougham's claim, in July 1812, that the lot of the Whigs was virtue and unpopularity, and in Sir James Mackintosh's belief that Whigs experienced a life of troublesome opposition with little chance of office and small hope of curbing abuses.[76] The trope of embattled opposition continued into the era of Reform government.[77] Whigs remained engrossed with the canonicity of their political traditions, even where this echoed despair, as with Lord Clarendon's exasperation in the wake of Whig failure to form a government in December 1845; the Whig party then, according to Clarendon, was both a matter of history and an effete relic.[78]

Expressions of pessimism emanated from every section of the Whig alliance at various times during the half-century in which it had a recognisably discrete continuity (c. 1794–1852). In 1798, John Russell, later sixth Duke of Bedford, wrote to George Tierney as the latter quit the Foxites, lamenting the annihilation of the Whigs following their parliamentary secession.[79] For Francis Horner, writing to Francis Jeffrey in September 1806, Charles James Fox's party had died with Fox.[80] The future lay rather with a popular party representing those with modest rising incomes who valued education and virtue.[81] In February 1835, the Whigs concluded their tentative alliance with the Radicals and O'Connellites against Peel's minority administration (the Lichfield House Compact). Yet in that month Sydney Smith agreed with Lady Grey that it was difficult to find among the Whigs any who were ready to govern.[82] In 1839, Lord John Russell opined that he had always thought that parliamentary reform would destroy the Whig party.[83] The narrative of enervation is a powerful one, because it can be told in the words of the Whigs themselves.

Now this in itself should inspire caution. Why were exclusion from power, political hypochondria and paranoia about the King such persistent themes? One answer, of course, is that when Whigs prognosticated their own collapse, they merely reflected the real and substantial difficulties of Whig politics. Another reason, quite possibly coexistent with the first, was that this type of rhetoric was functional to a transcending conceit: the notion that Whig statesmen were in political life despite their better interests. In 1827, the Fitzwilliamite MP James Scarlett had likened the Whig habit of refusing office to a man who would die pining for a fruit with his mouth open under a tree rather than clamber up to get it.[84] The context of this remark was Grey's scrupulosity about joining Canning. It was good political and rhetorical manners for Whigs to show that office was *not* their single object, even if this was really rather hypocritical. It is worth observing this mixture of pessimism and renunciation, both because it was a long-standing and commonly shared pose and because natural philosophy had an obvious utility in showing how Whigs were capable of transcending politics.

To give an example, the plot setting for Henry Brougham's *Dialogues on Instinct* was the general election of 1837, when 'every day brought the accounts of those mighty boasts of our expected successes under the new reign [of Victoria], so idly made, being overthrown by the activity and resources of our adversaries and the listlessness of the people on our behalf'.[85] While Brougham was out of favour with the Whig leadership, he demonstrated the etiquette that the pursuit of place should be disclaimed. Brougham's exchange of 'sound philosophy' with his interlocutor Lord Althorp gave him a lofty seat from which to criticise 'the liberal party, among whom the single object seemed to be the retention of office'.[86] The *Dialogues* are as much about Brougham as about instinct and this aspect has been underplayed by science historians concerned to construct an intellectual genealogy for Charles Darwin or to explore entomological debates.[87]

In the *Dialogues*, Brougham chose two Ciceronian texts as his models, namely *The Republic* and *The Laws*. The dialogues open with the characters Brougham and Althorp wandering in an arcadian idyll: the lakes, rivers and woods of Westmoreland.[88] Coming to an island in the fork of a river, resembling the Fibrenius, Brougham begins the discussion by quoting the appropriate depiction in Book II of the *Laws*. Both out of office, the one by retirement and the other by prohibition, Althorp and

Brougham elaborately affect their weariness with affairs. At the same time, reflecting on Whig losses in the 1837 election, Brougham again quotes Cicero, this time from the *Republic*: 'By our own misconduct, not by any calamity, though we may still have the name of a free government, we yet have lost the reality.'[89] The withdrawal from politics into speculation is itself a criticism of the regime that had dispensed with Brougham's services and lost political ground. The Ciceronic statesman waits for his services to be summoned once again. The point here is not chiefly the actual circumstances of Brougham's marginalisation, but the style of self-presentation. The withdrawal occurs within a landscape that is modelled on a classical text. Brougham's use of science manages to be both political comment and detached escapism. The knowledge of the statesman not only relates to his ability to comprehend affairs but is also a mode of disinterest. Withdrawn from 'little politics', the liberal statesman can survey the field anew. There is an Olympian conceit here, as well as pure diversion. This kind of narrative was consistent with progressivism, but it had its own dynamic and purpose too.

Whiggery and the spectrum of science and politics

The history of science has adopted a number of strategies to place really-existing Whiggery within its broader intellectual setting. A picture of Whig activity tends to emerge out of discussions of a range of different frameworks, which are the primary topics of analysis. These can be biographies, but are more often studies of particular learned societies, informal networks, urban civic locations, ministerial policies and patronage, or large-scale cross-disciplinary institutions such as the British Association. To focus directly on Whig manners is to cut across these categories to some extent. It also forfeits some of the social depth that is available to the history of scientific structures. Nevertheless, it can provide some help in coordinating the various historiographic currents.

The complexity of Whig positions described in the previous section has implications for discussions in the history of science. Whiggism has been described as a gradualist progressive response to competing pressures for political fixity or adaptation, in which the past is assimilated teleologically to the present.[90] This view owes much to Sir Herbert Butterfield.[91] A more sophisticated tack is taken by Richard Yeo, who strives to assimilate the nuance of more recent historiography. Yeo has recognised that whig progressivism sought to recapture the past as moral index to the present, as much as to orientate the past to a teleological present.[92] One should also state that Foxite inheritance and the breadth of Whig opinion went beyond a linear spectrum of progress and conservatism.

Norton Wise and Crosbie Smith identify a progressive, improving and liberal scientific culture.[93] This included moderate conservatives and philosophic radicals (such as William Whewell and John Stuart Mill respectively on the right and left), but it was centrally and typically Whig. An example of a figure standing at the centre is John Herschel (1792–1871), who touted an unabashed Whig view of history: a view shared by his Cambridge connections, such as William Whewell (1794–1866) and George Airy (1801–1892). However, much of the debate over the operation of a Cambridge network identified by Susan Cannon has centred on the plausibility of congruent mapping of individual religious, social and political positions.[94] Pietro Corsi has shown that Oxford Noetics had greater affinity with Dugald Stewart's epistemology than did Whewell.[95] While it is true that John Herschel was eventually regarded as embodying (as James Secord suggests) the magisterial adjudication of scientific common sense,[96] it is not clear if he was consistently regarded as a Whig. Separately, cosmic progressivism was associated with doctrines that he treated with scepticism, such as the nebular hypothesis explored by Simon Schaffer.[97]

The problem of defining the centre is not confined to the Cambridge network. In his seminal study of the *Politics of Evolution,* focusing on London milieus, Adrian Desmond explains that the focus of the book moves from left to right. The series roughly runs from anti-corporation radicals, Dissenters and unprivileged private teachers, through to the Whig moderates and then on to the wealthy anti-Lamarckians of the Royal College of Surgeons.[98] The moderate Whigs are represented by a gentlemanly figure such as Sir Charles Bell (1774–1842), who was educated in Edinburgh and was an ally of Henry Brougham in advancing William Paley's natural theology.[99] Within London University, the Paleyite Whigs were curbed by medical radicals who combined an attachment to progressive transmutationism with opposition to the anti-working-class policies of the Whig Reform governments, such as the Poor Law Amendment Act of 1834.[100]

While Desmond's coordinates are navigable in the world of metropolitan medicine and morphology, they appear to come adrift when transposed to wider political spaces. Thus, in a London University context, the Broughamite synthesis of utilitarian reform, political economy and social containment through secular education can plausibly be described as moderately centrist. But Henry Brougham was frequently viewed with distrust by his aristocratic Foxite parliamentary colleagues for whom he represented unpredictable extremism. William Thomas' distinction between the high Whiggism of the dynasties and the rank-and-file Whiggism of Brougham and his associates hints at the suspicion with which Brougham was often regarded by his contemporaries.[101] The problem was Brougham's erratic populism and flirtation with the democratic arguments of philosophic radicalism. Yet at the same time, according to Peter Mandler's argument, those aristocrats and gentry politicians closest to the Foxite tradition, c.1800–1830, regarded the apparatus of liberal political economy, including Poor Law reform, with a decided lack of enthusiasm. As explored below, the physiology patronised at Holland House was closer to the Francophile materialism of the medical radicals than to Broughamite dualism. Does this mean that aristocratic Foxites were somewhat to the left of the Broughamites, in modern perspective? Economically and scientifically they were; constitutionally and socially they were not. The axis of left and right readily begins to break down between discrete spectrums of status and opinion.

In the context of this kind of general problem, it is not surprising that historians have had difficulties in placing Whiggery in relation to Peelite conservatism and natural science in the 1830s and 1840s. Susan Faye Cannon asserts that Peel's conservatism was politically liberal, and that its true opposite was Whiggism. Peelite economy contrasted with the willingness of Prime Minister Melbourne (1835–1841) and his Chancellor of the Exchequer Thomas Spring Rice to grant money for scientific projects such as the Ross Antarctic Expedition and the building of magnetic apparatus.[102] Yet Roy Macleod, for whom Melbourne's administration was weakly liberal, argues that his government was inactive in fostering science.[103] The overhaul of the Royal Society between 1830 and 1848 was not a victory for the Whig progressive vocational agenda advanced by Charles Babbage (1792–1871) and his allies. Instead it was an achievement of a conservative approach to coalition-building and compromise.[104] Both Cannon and Macleod associate Whiggism, historiographically and politically, with a progressive professionalisation narrative that they are both concerned to revise.[105] Within this shared project, both authors reach contrasting conclusions about Whig proactivity in patronising science by drawing upon different types of evidence. Cannon sees generous responses to requests for capital spending; Macleod searches in vain for civil list pensions for scientists. Yet the problem is ameliorated if the sensitivity towards civil list expenditures inherited from Rockinghamite rhetoric is recognised. It was consistent for Whig governments acting in this tradition to limit personal patronages under the civil list, while encouraging spending on other initiatives. So far as the Royal Society is concerned, the guiding sensibility of leading Whig aristocrats was not particularly professional. In fact, coalition, conservatism and compromise describe it rather well. Reform could be whiggish, in this sense, and still sidestep Babbage's polemical vocationalism.[106]

Whig politics and the history of science

L.S. Jacyna has called for a unified study of the history of science, political theory and politics.[107] This aim is admirable, and the present argument seeks to tackle it, but it has to be recognised that these genres have distinct explanatory focuses that go beyond their different subject areas. Contemporary historians of science have inherited a fundamental debate about the nature of scientific epistemology which is still not exhausted. The idea that science has expressed wider sets of ideological commitments has been particularly useful to historians seeking to escape internalist accounts of scientific concepts and disciplines (although the internalist/externalist framework has now really been superseded). Over the past three decades, many historians of natural science have incorporated political contexts into their accounts of the making of natural knowledge.[108] They have given weight to the politicised behaviour of individuals within scientific environments and to political motivations in the adoption of scientific ideas in policy.[109] We shall take a closer look at the purely theoretical issues in the Conclusion (including the disentangling of Whig science as an analytical category from the actual science of historical Whigs), but the general point to note here is that the agenda of historians of science is distinct from that of political historians, even as they survey much the same ideological and cultural source material.

Partly because of this difference, historians of science have been much more sensitive to sociological and anthropological approaches than their political counterparts. An outstanding example of this sort of integration with an eye on the fundamental questions of truth-making has been furnished by Steven Shapin.[110] While Shapin focuses on the early modern period, his appreciation of the role of civility in building up relations and reputations of trust among gentlemen can be emulated in the study of Whig manners. A line can be traced between Shapin's Restoration courtesy and the gentlemen of the British Association identified by Jack Morrell and Arnold Thackray (although it should be noted that Morrell and Thackray do not concentrate on the construction of a truth-community in Shapin's terms). James Secord had drawn an explicit parallel between the gentility identified by Shapin and the codes of independent honour underpinning early-nineteenth-century science, while stressing the gulf between these codes and the etiquettes of modern research.[111] The political and scientific world of the early nineteenth century comprised overlapping communities with different demands.

We have already mentioned the political contraction of Whiggery in the latter part of the reign of George III. At the same time, there was an important sense in which Foxite Whiggery remained central to the political scene as the conceivable 'other' or alternative power network. It is true that revolutionary forces (especially after 1790) threatened the whole system from without, and that various streams of radicalism began to make their presence felt in the early years of the nineteenth century. But any feasible parliamentary alternative demanded the inclusion of the Whig opposition, much as the Talents ministry did in 1806, or the Canning-Goderich governments in 1827–1828. Now in contrast, Whiggery had no obvious *intellectual* right of place in the much broader argument about theology, politics and nature that raged in the aftermath of the French Revolution. After the conflicts earlier in the century, the hegemonic Whig world-view of the mid-eighteenth century (with its Newtonian and latitudinarian biases) was in many ways ill-equipped to deal with the new era of revolutionary polarisation. In 1753 Horace Walpole could poke fun at the decline of Methodist fortunes in Oxford and the vogue for Hutchinsonianism that had replaced it.[112] By the end of the century, such Whig complacency was outmoded in a new and more desperate landscape of ideological struggle. J.F.M. Clark describes how (in the context of natural history) atheism and materialism stalked the established order.[113] Whig statesmen and intellectuals could no longer afford condescension towards philosophical debate. When Foxite Whigs and their allies attempted to redefine and reassert their natural philosophic position – with some success in the case of the Edinburgh reviewers – they were doing so from a station of weakness, and were essentially racing to catch up with a debate whose terms were set by other forces.

This can be seen even in the strongest sector of Whig intellectual life, north of the border. The intellectual ferment in the Scottish universities, with their strong Whiggish and civic republican heritage, and the emergence of a knot of pro-Foxite intellectuals, should not distract us from the fact that the French Revolution was bitterly divisive in

Scotland.[114] Many Scottish natural philosophers were attached to the system of Pitt and Dundas, and their tory government successors, and violently opposed anything that smacked of opposition. Morrell has described how John Robison produced a particularly virulent attack on revolutionary freethinking and freemasonry, while the Whig professors Dugald Stewart and Playfair acted cautiously to preserve their intellectual discretion, and only raised their heads politically under provocation from the Moderate faction inside the Kirk.[115]

With the partial exception of Scotland, Whigs were not the prime movers in the development of science. To begin with, the leading English natural philosophers, technologists and experimentalists were not usually Whig partisans. The Birmingham and Lichfield Lunar Men perhaps came closest in sympathy, yet they could not really be described as Foxites in a party sense. Priestley had a previous connection with Fox's new and uneasy ally, the first Marquess of Lansdowne (Lord Shelburne). The pro-Jacobin radicalism of the younger generation contrasted with the Anglican pragmatism and humane enquiry of the elder Watt and Boulton.[116] Erasmus Darwin (1731–1802) was strongly opposed to slavery, but the Foxites did not have a monopoly over this cause. The northern civic Enlightenment (signally at the Liverpool and Manchester Literary and Philosophical Societies) had reforming tendencies, but these were only patchily integrated into Foxite politics and were stronger among the rank and file than figures of national prominence. The Mancunian John Dalton (1766–1844) eventually benefited from patronage under Grey's government, and pressed for scientific reform through the British Association. However, he had few connections with metropolitan or rural Foxite circles during the main body of his career. Indeed, in 1826 Dalton received the first royal medal from the Royal Society, courtesy of George IV. In the next generation senior figures such as David Brewster (1781–1868), a Scottish friend of Brougham's, Charles Babbage (1791–1871) and Adam Sedgwick (1785–1873) can be counted as Whigs, but again the political centre of gravity among many other influential intellectuals in the 1830s tended to be tory. Susan Cannon's Cambridge network divided fairly evenly between Whigs and Peelites such as William Whewell whose Liberal Toryism was of the Malthusian, anti-deductivist type.[117]

Lawrence Goldman has drawn attention to the way in which Whig noblemen and MPs came forward to patronise the new science of statistics in the 1830s without properly understanding the anti-Ricardian, inductivist agenda of the intellectuals who formulated it.[118] The theoreticians Richard Jones and Whewell were concerned that the profile of the Statistical Society in 1834 fitted too closely with the Whig government side.[119] Meanwhile, Fitzwilliam, Russell and the other Whig politicians were more interested in the use of data for social improvement than in overturning orthodox political economy. Their intentions had more in common with the provincial statistical societies, such as that in Manchester, which were concerned with the 'Condition of England' debate.[120] Mary Poovey has argued that the various parties to the theoretical debate over induction bracketed or foreclosed their disagreement by opposing mathematical method to rhetoric.[121] But while this rings true of the theoretical participants, it did not have much practical effect on Whig politicians. Whig statesmen wanted to have it both ways: not only to have statistical material but also to manoeuvre rhetorically. The division of labour usually operated so that patrons could sponsor intellectual consultants at discretion.

Whigs were actually far from dominant in the structures that organised British science after 1780, although they did become more influential from roughly the mid-1820s onwards. It was not that they were completely excluded, but that leadership passed into other hands with other agendas. The shift at the Royal Society was probably most significant, because Sir Joseph Banks used his long presidency (1778–1820) to entrench a powerful network of patronage and institutions that was clearly aligned with the Pittite ascendancy.[122] Before his opposition alignment was firmly set, Brougham wrote to Banks (who had known his father) asking him to consider 'hinting something in my behalf' in political circles – a tribute to Banks' influence. 'All this, however, resulted in nothing.'[123] A degree of personal affability and civility encouraged Whigs to try to court him: John Gascoigne indicates that Banks remained on good terms with Fox, who paid tribute to him.[124] As Richard Drayton recognises, the Bedford connection played an important part in keeping channels open between the opposition and Banks' learned empire.[125] However, there were limits to this cordiality – Banks had distant relations with Holland House.[126] Banks' brand of rational, patriotic imperialism was differently oriented to the civic republican patriotism characterising Foxite Whiggery.

Banksian politics was inescapably global in perspective.[127] Banks was able to operate a world network of botanical surveys and plant transfers through Kew Gardens with the cooperation of Pitt and Dundas.[128] Banksian allies included Arthur Young and Sir John Sinclair at the Board of Agriculture. Banks and Sinclair lobbied for agricultural protectionism, and for agrarian investment, if necessary at the expense of the Sinking Fund.[129] C.A. Bayly associates such agrarian patriotism with the first phase of imperial state reform under Pitt. Self-sufficiency in food would provide a secure basis for the projection of national power.[130] While not necessitating territorial acquisition, British power could then be used to secure captive markets.[131] Banks' enlightened mercantilism dovetailed both with George III's conception of patriotic kingship and with an era of intensifying imperial competition. If anything, the loss of the American colonies reinforced the idea of domestic improvement as the basis for imperial reassertion.[132]

Whig members of the landed elite could certainly participate in the chemical and agricultural structures associated with Banks' general purview, through the Board of Agriculture and Royal Institution. We will discuss this at greater length when we come to Whig uses of the georgic tradition in Chapter 5. This Whig agrarianism was compatible with the comparative lack of interest in expanding the commercial empire shown by opposition Whigs before 1815. A devotion to peace and liberty abroad was combined with a cultural internationalism in Foxite francophilia (anticipated by Rockinghamite sympathy for the revolutionary American colonists), but not with the same emphasis on systematic imperial exploration and exploitation as sponsored by Banks. The Royal Society and Royal Institution could easily be cast as courtly bodies. Drayton observes that when Banks was made Knight of the Bath (1795) and appointed to the Privy Council (1797) he was lampooned as a base fawner.[133] Brougham developed the theme of antagonism to courtly metropolitanism in his *Edinburgh Review* articles on Young and Davy.[134] Younger Whigs associated with newer entities that were more independent of Banks' learned empire such as the Geological Society. This is not of course the same as Whig boycott or exclusion, and a figure such as Humphry Davy could move quite easily between the worlds of Banksian institutional and Whig aristocratic sponsorship. Nevertheless, relative Whig marginalisation in these years might have been one long-range reason for Whig patronage of the new organisations such as the British Association that arose in the 1830s in the wake of the 'decline of science' debate (although it was never framed in such crude terms as a Whig *revanche* and Whigs did not monopolise the new bodies).[135] During the 1830s Whigs moved more decisively towards the centre of scientific patronage, partly as a function of their increased political power and confidence. Drayton explores this Whig renaissance in botany and agrarianism, noticing that many of the leading figures were the same Young Whigs identified by Wasson.[136] Of course there had been continuing Whig involvement through the Bedford clan, but Drayton seems right to claim Whig consolidation.

By the beginning of Russell's first government (1846), British politicians from across the spectrum were well entrenched as figureheads of British learning, including many Whig statesmen. Their effective participation was occasionally considerable, but more important than this was their visible patronage and support. In 1847 the Reverend Abraham Hume published his survey of *The Learned Societies and Printing Clubs of the United Kingdom*, claiming that 'the references to LEARNED SOCIETIES are so frequent, in our age, in books, newspapers, conversations etc. that every man who claims to be educated is expected to possess some knowledge of the subject'.[137] Hume's survey was extensive, if by no means exhaustive. A review of the members of parliament and peers listed among the figurehead roles (chiefly presidents and vice-presidents) of Hume's societies gives a very rough snapshot of the distribution of political patronage. Being a figurehead was an example of public style shared among the elite, although it could signify very different things in different cases. Still, when one tallies the 66 MPs and secular peers who were presidents or vice-presidents, 17 could be classed as Whigs, 19 as liberal tory or Peelite, and 11 as high tory. The rest were either based abroad, impossible to place even approximately in any camp, non-political, or did not leave much indication as to their political inclinations.

While no great emphasis should be placed on the exact numbers, the fact that there were half to two-thirds more Whigs and Peelites again as high tories, and a roughly even number of Whigs and Peelites, stimulates further questions about the distribution of statesmen across societies. Thus we find the Whigs concentrated in the Horticultural Society (the Duke of Devonshire, the Earl of Auckland, and Sir Charles Lemon, MP for Cornwall); the Royal Botanic Society (the late twelfth, followed by the thirteenth Duke of Norfolk, the Earl of Albemarle), the Statistical Society of London (Earl Fitzwilliam, Lemon); the Archaeological Institute (Fitzwilliam and Lemon); the Sussex Archaeological Society (the thirteenth Duke of Norfolk); the Royal College of Chemistry (the Marquesses of Lansdowne and Breadalbane, Fitzwilliam, the Earl of Clarendon, Lords Brougham and Russell, Lord Grosvenor, Lord Palmerston – at this point Whig-aligned – and Sir Charles Gray); the Royal Geological Society of Cornwall and Royal Cornwall Polytechnic Society (Lemon); the Yorkshire Philosophical Society and Geological Society of the West Riding (Fitzwilliam); the Literary and Philosophical Society of Whitby (the Marquess of Normanby) and the Literary and Antiquarian Society of Perth (Breadalbane and Francis Stuart, the fifteenth Lord Gray). Hume's list confirms Drayton's observation that

Whigs had made progress during the 1830s in penetrating the commanding heights of botany, horticulture and agriculture. Whigs had kept their foothold in the statistical and geological world, and in local strongholds. They had been recruited to the new college of chemistry. But what was the nature of the political manners exhibited in such roles?

2
The Statesman

The knowledge of the statesman

The intensification of high political interest in science in the last decades of the eighteenth century marked a changed pattern of self-depiction among politicians. In this, it mirrored the development, and indeed accommodation, of the notion of the 'statesman' itself. The rise of statesmanship dignified the politics of knowledge. The statesman flaunted his knowledge as a sign of political definition and display. The process could involve a detailed grasp of theoretical questions, but did not necessarily need to do so.

Recognising the changing treatment of statesmanship is all the more important because the late-Hanoverian and Victorian cult of the statesman was so pervasive that historians rather take its development for granted.[1] It became the stuff of dramatic tableaux: for example of Macaulay's stirring account of the death of Chatham in the Commons. As Macaulay noted, Chatham's spot in Westminster Abbey now stood for statesmen just as the other end of the transept was reserved for poets.[2] Another sign of the growing status of statesmanship was its incorporation into mid-nineteenth-century English political theory. J.S. Mill's preferred bicameral upper house in *Representative Government* (1861) was a Chamber of Statesmen, while Walter Bagehot wrote lyrically on the guidance and benefit to the nation of a statesman speaking 'in a palpable way [of] what is to him important truth'.[3]

It is worth emphasising, however, that this constituted a definite alteration in register. In the seventeenth century the statesman was occasionally referred to neutrally, or positively, as a servitor of public authority, but was more often bracketed in a hostile manner with the manipulative politician. In his depiction of archetypes in 'The World', the Royalist poet Henry Vaughan (1621–1695) drew a lurid picture of blood, tears and darkness.[4] The word 'statesman' is not used at all in Hobbes' *Leviathan* or Locke's *Treatises*.[5] 'Public minister', 'legislator', 'counsellor' and 'magistrate' are preferred technical terms with varied meanings in each system.[6] When Locke wanted to ridicule Sir Robert Filmer for traducing Homer and Aristotle, he called him a zealous and Christian politician.[7] However, to the aspiring statesmen of the 1820s, these were distant authors, whose subversive potential was contained by their remoteness. As one correspondent wrote to Lord Lansdowne, Locke was harmless so long as one read him knowing that he was controverting Filmer.[8] However, it would be wrong to dismiss this as the disdainful comment of someone who had no interest in or understanding of seventeenth-century controversies, for the same person considered the aristocratic republican Algernon Sydney as intellectually a living figure.[9] Rather, Hobbes and Locke were the partisans of exploded controversies, while Sidney stood at the head of a recognisable eighteenth-century genealogy of civil aristocratic power.

Yet the road from the aristocratic patriot to the virtuous public statesman was far from straight. The evidence points to transition in the later eighteenth century, and Harvey Mansfield is surely right to point to the pivotal part played by Edmund Burke's polemic against Bolingbroke, and exoneration of party in the 1770s. But Mansfield mistakes the character of the change because he assumes that statesmanship had a stable positive signification (in effect, reading back the post-Burkean definition) and was therefore an arena of conceptual competition.[10] On the contrary, Bolingbroke's patriotic politics represented an attempt to escape statesmen and their statecraft, while Burke's prudential statesmanship was a new construction, forged in the specific conditions of George III's hostility to various connections among the Old Corps Whigs. As late as 1755, Samuel Johnson was defining the statesman as simply 'a politician: one versed in arts of government', with cynical illustrations from Pope and Addison referring to the management of corruptibility.[11] In his *Essay on the History of Civil Society* (1767), Adam Ferguson (1723–1816), synthesiser of civic republicanism and Scottish philosophic history, retained the older prejudicial characterisation of the politician-statesman, and gave it a newer disturbing implication. The function of the statesman was created by the progressive division of labour, so that the statesman increasingly acted blindly, like the part of an engine.[12]

Ferguson's reference to the cognition of the statesman was a notable shift. In contrast, the trustworthiness or otherwise of political rhetoric was a problem of political thought going back to the early modern period.[13] Writing contemporaneously with Ferguson, Sir James Steuart made the knowledge of the statesman foundational to his political economy.[14] Its importance was also evident in *The Wealth of Nations* (1776). When Adam Smith (1723–1790) sought to uphold individual direction over the employment of capital for production, it was the deficient knowledge of the statesman that concentrated his fire. The individual could deploy his own capital better than a statesman could for him. The attempt to control a myriad of private circumstances would overburden the statesman, representing an arrogation of authority that was bad enough in a deliberative body and worse in an individual deluded enough to think that he could govern thus effectively.[15] There was disgrace attached to lack of knowledge and the misuse of authority that arose from it. Of course, such interference was also unjustified because it impeded the virtuous independence of judgement, and consequent interdependence of moral agents in the exchange economy. However, it seems clear that Smith closely related the character of statesmanship to its state of knowledge. This is illustrated by his observation that in barbarous societies every man acts as his own statesman and is able to form competent judgements about public interests because of the relative simplicity of occupations and the alertness of mind prompted by practical challenges.[16] There is a rough equality of knowledge. In general, the extent of the statesman's activity is very largely justified in proportion to his apprehension. Donald Winch cites Smith's contemptuous reference to the craftiness of the political animal at the mercy of fluctuating pressures of interest.[17] Yet the context of the passage makes it clear that this description is a grudging recommendation. Where commercial and political circumstances fluctuate rapidly (for example in trade disputes) the craft of the statesman or politician might be more useful than legislative science operating by general principle.[18] The animal has his uses, and there are certain situations in which the judgement of varying conditions is apposite. If this is an attack on politics, then it is a qualified criticism.[19] Smith maintains the division between the scientific, generalised knowledge of the legislator and the artful, contingent knowledge of the politician-statesman. Smith's statesmanship is therefore not free of the earlier pejorative connotations of manipulation, but the cognitive element is decisive in the argument.

Subsequently, the heroic transformation of the statesman gathered pace. In comparison with the politician, who retained the faintly discreditable implication of faction, the statesman came to embody a judicious mode of cognition, balancing general principles and particular circumstances for the good of his country. That is, the legislator's science was merged with the statesman's art. As Edmund Burke put it,

> A statesman differs from a professor in an university; the latter has only the general view of society; the former, the statesman, has a number of circumstances to combine with those general ideas, and to take into his consideration. Circumstances are infinite, are infinitely combined; are variable and transient; he who does not take them into consideration, is not erroneous, but stark mad – *dat operam ut cum ratione insaniat* – he is metaphysically mad. A statesman, never losing sight of principles, is to be guided by circumstances; and judging contrary to the exigencies of the moment, he may ruin his country for ever.[20]

The tension between principle and the management of circumstances made statesmanship rhetorically vulnerable: on the one hand, to charges of inflexible abstraction; and on the other, to imputations of self-interest and transient manoeuvre. But the demands of knowledge also made statesmanship powerful as an operative ideal, so long as successful and transcending comprehension could be shown. This precariousness was manifested in real political battles and wars of rhetorical manoeuvre. During the general election of October 1812, in Liverpool, George Canning described his Whig opponents (Henry Brougham and Thomas Creevey) as speculators and mischievous reasoners.[21] One leading local Whig, William Roscoe, retorted,

> On the contrary it must appear to every intelligent and impartial reader that these harangues exhibit nothing of those strong feelings for the happiness of the people, those comprehensive views of the public interest, or that deliberative wisdom which ought to characterise even the most extemporaneous effusions of a truly enlightened statesman.[22]

Canning was a vulgar orator unable to see the public interest or to deliberate upon it. Meanwhile, the enlightenment of Brougham and Creevey was displayed in frequent Whig demonstrations at the Liverpool Botanical Gardens, co-founded by Roscoe in 1802.[23] The self-depiction of reflective knowledge was a potent weapon in the hands of Whig–Liberals fighting off various challenges; moreover, it was a style of behaviour that was to be learnt by fledgling politicians, as illustrated by the following incident.

In 1822 the young Whig MP Edward Stanley (1799–1869) wrote to Lansdowne to ask for guidance in his historical and political reading so that starting systematically he might avoid disgrace in public life.[24] This was not the same as asking for ideological instruction, and was not received as such; Lansdowne's long and detailed reply focused on models of argument and eloquence, and sources of information on the opposing sides of various public questions. It was an elegant piece of social courtship, by a junior of a senior politician, and it elicited an invitation to Stanley from Lansdowne, suggesting that he come down to Bowood (Lansdowne's Wiltshire seat) in order to consult the proposed works there. Stanley's alignment with Lansdowne was subsequently demonstrated in 1827, when Lansdowne procured for him the post of Under-Secretary for the Colonies in Goderich's coalition government.[25]

Lansdowne's advice to Stanley took the form of presenting him with a canon of notable works and a lengthy commentary on this canon.[26] What is particularly striking is Lansdowne's explicit admission that his was a list of books with which he had not acquainted himself as much as he ought, though he was confident of their value. In other words, the list was not just a repository of ideas, but a normative artefact in itself. Showing knowledge of the authors' texts had value alongside acquaintance with their contents. They embodied the sort of thing, as a rule, that the statesman ought to know. In other words, it functioned as a signal. It was perfectly possible to brandish the books without completely knowing what was inside them. Subsequent remarks reveal which works were more or less familiar to the marquess. For example, Lansdowne admitted that he had not yet read Lingard's incomplete *History of England*, but expected it to be accurately researched, and in any case it should be read to give a Catholic point of view. It was incumbent upon the statesman to gain such rounded knowledge. The recommendation of Burke is vague enough, although special interest in Burke's earlier writing is indicated by attribution to him of the historical section of the Annual Register dealing with the first years of the reign of George III (1760–1820). Lansdowne was able to specify that the second volume of Dugald Stewart's *Philosophy of the Human Mind* contained the best investigation of logic that Lansdowne had encountered. Together with other works of metaphysical science, including Francis Bacon's *Advancement of Learning* (recommended over the more popular essays) and the works of Bishop Berkeley, they constituted the best models of style and eloquence. In a similar vein, Stanley was urged to read Adam Smith's *Wealth of Nations* (though Lansdowne assumed that Stanley was already familiar with it). Not only was it the basis for

Malthus and Ricardo, but it was to be admired for its clarity and style of reasoning.

Several preoccupations are evident throughout these observations. The first is the need of the statesman to resist vulgarised oratory and newspaper writing, which is a form of corruption.[27] Eloquence (the proper use of style to convince by appealing to reason) is upheld against oratory (cast here as the gateway to demagoguery, appealing to the passions). It can be learnt not only by emulating the ancient classics, Machiavelli's discourses on Livy, Bayle's Dictionary, Erskine's speeches, the French masters of expression (Fenelon, Bossuet, Rousseau and Montesquieu), but also by absorbing lesser works – anonymous pamphlets concerning the education of the Irish peasantry, for example, or tracts on the Neapolitan revolution. The second preoccupation is the statesman's obligation to absorb and synthesise large amounts of legal and historical information. A third theme is the advantage in cultivating the capacity to formulate balanced arguments on contentious issues: the neutrality of shipping, for example, or the state of Ireland. All of these are essential attributes for a statesman. Lansdowne's emphasis on statesmanship is critical in describing the public life that Stanley wished to fit himself for, and distinguish himself in. Political signalling was a process of *referring* to one's possession of and connection with knowledge, rather than necessarily enacting or performing it. Patronage – in this case Lansdowne's offer of assistance – was itself a powerful form of signal. For instance, it could refer to the deferred distribution of goods (Lansdowne and Stanley were stocking up mutual goodwill). The aspiring statesman had to be in a position to signal his awareness of allies, and his appreciation of the statesman's role, its accoutrements and expectations.

There was a spatial aspect to the making of style too, which should be noticed: the self-portrayal of cogitation and reading within physical imaginative surroundings. In effect, this was the use of locale to depict Whig identity. The phenomenon is instanced in Francis Horner's sojourn at Torquay in 1811. With the help of his brother, Horner hoped to trace the relations 'which the slate, the limestone, and the horizontal sandstone mentioned by Mr. Playfair, have to one another'.[28] This locale becomes a location of ideological or aesthetic identity with the claim that 'This sea, with its beautiful shores, and the neighbouring mountain…explain Homer better than a score of scholiasts.' In a Romantic trope, scientific wandering in nature becomes authentic literary criticism. Torquay is a place of meditation, of escape from 'little politics'. Yet Horner confesses that has brought along Playfair's

Illustrations and Paley's *Natural Theology* to accompany the chief business of reading 'Machiavel's Discourses on Livy , Montesquieu's Greatness and Decline of the Romans, Hume's Political Discourses and Burke's Tracts on the French Revolution'.[29] Horner wishes to bring all these texts together 'so as to compare them, and make them as it were sit in council, in my hearing upon the same points'. His 'purpose in studying them is to apply their reasonings to the awful and desperate circumstances of our own time, and to apply these circumstances to their reasonings'.[30] Horner's reading list is rather similar to Lansdowne's. The fact that Machiavelli's *Discourses* appears in both provides some corroboration for the hypothesis that neo-roman argument continued to be influential into the nineteenth century. More importantly, the locale of geological Torquay merges into a literary display of broad reason.

Instances of scientific display were not only literary in character. When Charles Lyell took the Chancellor of the Exchequer (Charles Grant, not Charles Wood) and the President of the Board of Trade (Lord Clarendon) on their oolitic excursion from Lansdowne's seat at Bowood, it was because the ministers 'wanted to forget the state of Ireland, growing as it does every day worse and worse' (December 1846).[31] The profundities of geological time also served to distance the acuteness of the Irish disaster, or more precisely to provide uniformities against which the catastrophe could be put into a vaster temporal perspective. The third member of Lyell's geological party and another of Lansdowne's guests was the liberal Anglican historian Henry Hart Milman (1791–1868). On a previous evening, just before their trip to view the Bath oolite, Clarendon, Milman and Lyell had discussed the antiquity of historical time with Milman noting the discrepancies between the various biblical chronologies (Hebrew, Septuagint, Samaritan, Mosaic and Pauline). Clarendon had concluded, 'It seems then, that the geologists are blamed for not making their notions of the world's age agree exactly with a chronology (supposed to be scriptural) which we really know nothing about.'[32] The President of the Board of Trade had not in Lyell's opinion 'read or thought much on the point before', but he proved to be an 'apt scholar' in the field.[33] The geologist also commended Clarendon for his 'liberal' and 'statesmanlike' views on American commerce, surprising for one of his aristocratic station.[34] Another of Lyell's interlocutors at Bowood was the civil servant Charles Greville (1794–1865), whom Lyell praised for his 'excellent work on Ireland' (*The Past and Present Policy of England towards Ireland*, 1845).[35] This was a polemic against the privileges of the Church of

Ireland, and in favour of free trade.[36] Greville wished to discuss with Lyell the question of the comparative antiquity of the Cheirotherium and Megatherium.[37] Among this little knot of liberal-minded Whig public men it was the question of chronology that animated the geological conversation.

The participants in this sociable encounter did not really use the episode to lay out a tightly articulated theory of history and natural history. Instead they interacted in a particular large-minded way, signalling a kind of solidarity. The ideas expressed more than ideology. They were the currency of a form of sociable signal. Part of this style was a pose of cultivated detachment and scepticism, avoiding dramatic declarations of the apocalyptic or sublime. Of course there was an ideological dimension, and one can identify some of these elements with the sedimentary gradualism identified by John Burrow. The temporal framework of the conversation was ordered and expansive, rather than (in Burrow's terms) suffocating and claustrophobic.[38] To give another parallel, Duncan Forbes describes Milman's history as panoramic in character.[39] Milman's ascending cycles and analogous periods are depicted within physico-moral environments such as a manufacturing wilderness.[40] Lyell maintained a shrewd interest in Milman's teutonic latitudinarianism and in the clerical controversy that the latter ran into over his *History of the Jews* (1829).[41] Lyell's debt to Niebuhr has been well attested, and this observation is amplified by Lyell's evident monitoring of the parallel Niebuhrised historiography of Milman.[42] The principal point here is that in the close sociability of the Whig salons, Lyell's chronological reasoning was drawn upon in a similar way to Milman's. For Clarendon, scriptural geology was undermined through the combination of historical biblical study and Lyell's field lessons. The social experience at Bowood integrated episodes of comprehensive conversation with those of geological roaming. An irenic evocation of nature was a response to political turbulence and cataclysm. The geology enjoyed by Whig statesmen demonstrated lofty escape from political-economic catastrophe. It was a stylistic pose of disinterest, as well as a manifestation of the Whig ideology of history.

Administrative statesmanship

One variant of Whig politics was probabilistic, precautionary and tentative. Lansdowne and Lord Grenville were quintessential administrative liberals in this sense. They can be thought of as 'magisterial' because of their administrative focus and didactic slant; they wanted to compete and perhaps cooperate with Pittites and Liberal Tories to run the state.[43] Their approach tried to assume the most adverse conditions, and so can be thought of as a pessimistic kind of improvement. This pose was defined not only against *philosophe* rationalism but also against the opposition idioms of Foxite and country-radicalism. By contrast, Whig grandees most closely connected with the Fox–Holland House circle such as Holland and Grey managed to be aristocratic without being magisterial; that is, without being primarily focused on the arbitration and administration of public power. Essentially they defined themselves as checks upon the magistrate, rather than as potential administrators. Scots entering the political scene tended to orientate themselves along either the magisterial register (Horner) or the oppositionist register (Brougham), while confederating in the same party. The *Edinburgh Review* cohort had to find some way of fitting into this spectrum, and this proved to be as great a challenge as other more theoretical imperatives, such as extending the Reidian legacy.

Magisterial and oppositionist Whigs shared a common element of the Burkean inheritance. This was the standard by which the knowledge of the statesman determined his value and his role. Those Whig statesmen who were magisterial liberals extended the rhetoric of judging the temper of the people and the times to practical problems of handling knowledge. Precedence was to be given to economies of information rather than to the liberties of the subject. In this way an anti-commonwealth liberal style of politics emerged within the Whig party at the turn of the nineteenth century, fostered by statesmen semi-detached from the Foxite core.

There had already been an underlying shift during the eighteenth century in the state's management and handling of public knowledge, much of it ignored by mainstream Rockinghamite and Foxite Whiggery.[44] This development has been extensively documented by William Ashworth, Joanna Innes and others.[45] The gradual albeit uneven spread of mensuration and the changing pattern of socio-military discipline (as Ashworth puts it, from thanatocracy to accountocracy)[46] raise the possibility that the emergence of a magisterial liberal style among Whig statesmen at the turn of the nineteenth century was more a process of catching up with surrounding trends than an innovatory stroke.[47] In any case, it is important to remember the ramshackle character of many of the state's information-gathering processes. The administrative growth identified by Joanna Innes, William Ashworth and others did not mean that statesmen thought that their resources were adequate.

A striking example is furnished by Lansdowne's consultation with Nassau William Senior (1790–1864) over the enumeration of proprietors and their incomes in 1831. Scrutinising the Lincoln's Inn library, Senior told Lansdowne (at that point, Lord President of the Council) that he could not find documents showing the spread of proprietors' incomes, save for property tax and house tax accounts.[48] The issue under discussion was the relationship between property-holding (especially real estate) and income. Senior thought that the tax office might yield invaluable information. Imperfect hints might be gained from the land tax assessments and the returns of the house tax.[49] The real prize, however, was the property tax assessments on land which stood to give comprehensive details. The problem was that no one knew whether these assessments had been destroyed or not. Senior found a reference to measures taken in 1818 to destroy the property tax records except insofar as might be necessary in accounting for arrears, but could not find the paper. He reported a tip-off from George Pryme, Professor of Political Economy at Cambridge (1781–1868), that the gentlemen of Cambridgeshire had once settled a rating dispute by getting hold of the property tax assessment for the county, the business being handled by the chairman of the Quarter Sessions.[50] Senior was reduced to reminding Lansdowne that the Domesday Book and feudal laws would show the extent of past properties.[51] He confessed that he was a bad antiquarian, assuring Lansdowne that there were many such persons he could turn to for that kind of research.

The antiquarian whom Senior particularly recommended was Francis Palgrave (1788–1861), who had been involved in the presentation of public records.[52] There is no evidence in the Lansdowne papers that the marquess followed up the suggestion, although later correspondence with Palgrave shows that the matter of railway regulation was discussed between the two in the context of the law of the King's highway.[53] The point to be pressed is that the resources available to ministers were patchy, and that measurement and assessment were liable to be frustrated. The contingent and cautious approach discussed below comes into sharper relief in the light of these uncertainties. Growing awareness of the importance of public information, combined with consciousness of the sheer frailty of many of the structures of information gathering, contributed towards a style that in practice acknowledged its own constraints. On the other hand, administrative Whigs felt compelled to maintain the rhetoric of the possibility of precise knowledge.

An example of this impulse is provided by Lord Grenville's 'Fragments on Political Economy' (1826–1828), a long half-finished manuscript that, as its dedication and organisation into chapters suggest, was intended for eventual publication.[54] In touching on this source, Hilton observes that the inspiration behind it was Butlerian rather than Benthamite, finding the grounds of political-economic theorems in moral probability, rather than empiricism or Ricardian deductivism.[55] The point can be expanded. Grenville believed that precise knowledge could be gained from introspection, which gave similar precision to political truths as to discoveries in physics or in gravitation. Grenville did not assume that political-economic knowledge was identical in character to that of physics, but he did assert that an equivalent degree of confidence in basic axioms could be ascertained from introspective reasoning, which could then work outwards in degrees of certainty.[56] The problem was to account for the internecine debates between theologians, for example, and the analogous differences of political economists. In both political-economic and religious controversies, truth was discovered through moral evidence. So there was variation in the force of moral proof as different minds were prepared in different ways to receive such truths.[57] It was this that lay behind his declaration against the Sinking Fund, cited by Hilton, that the aim of legislation was not the multiplication of wealth but the psychological or moral reception of wealth.[58] The status of such knowledge as indubitable in basis might itself seem rather dubious, but Grenville insisted that the debates of philosophers could not shake confidence in the existence of virtue itself, or in the possibility of firm knowledge of moral principles and duties.[59] Moral probability gained its force from the premise that somewhere a set of absolute philosophic truths really did exist. The question was then of the best procedure to pursue probabilism and share knowledge.

Grenville's aggressively self-confident philosophical axioms contrasted with the cautious and provisional character of his propositions in political economy, but this imbalance was part and parcel of his theoretical understanding. In true Butlerian fashion, Grenville proclaimed that to judge probabilities and to act accordingly was the proper use of men's faculties. However, perfect consensus was not to be expected.[60] There is some evidence that the models of reasoning that Grenville initially had in mind were medicine and agriculture.[61] The progressive process of ascending approximation was like a spatial journey, in which the distinctive properties of physical or moral objects in creation became clearer, whereas previously they were seen only at a distance and in mass.[62] The bulk of the rest of the half-treatise was a fairly unremarkable compilation of historical and classical evidence within a scheme derivative from Smith. Yet the substance of the manuscript is less significant than the blend that it attempted to maintain between dogmatic and contingent knowledge. Grenville tried to enact an intellectual procedure for managing uncertainty, predicated on sure premises.

Republican science and the problem of the Foxite alignment

The magisterial statesmanship epitomised by someone like Grenville was not shared by the Foxite heart of the Whig party. On the contrary, Foxite constitutionalism was characterised by a fierce reassertion of country and commonwealth politics in the Revolutionary and Napoleonic period. As Pocock points out, the war was condemned in terms reminiscent of the Toryism of the first decade of the eighteenth century: a critique of patronage, permanent military establishments and national debt.[63] On the one hand, Foxite rhetoric celebrated the balanced constitution confirmed by the Glorious Revolution, and decried the expansion of the militarised executive as a preparation for absolute monarchy. On the other hand, Foxite whiggery elevated this danger from the throne above the more general confrontation between revolutionary democracy and European civilisation perceived by Edmund Burke. This was the burden of the polemic maintained by Fox's own *History of the early Part of the Reign of James II* (published posthumously in 1808) against Burke's later canon.[64] Against this background, it is significant that Jacyna has shown how the commonwealth rhetoric was as important in Scottish natural philosophy as the sociological progressivism of 'Scientific Whiggery'.

This is a very useful insight, as the alignment between the Foxite rump and significant parts of the Scottish intelligentsia was neither necessary nor inevitable. The canonical texts of Philosophical Whiggism were those of philosophic history, political economy and moral science. Jacyna has indicated the presence of older, civic republican motifs in the scientific and philosophical societies of 1790s Edinburgh.[65] On the one hand, chemistry provided empirical knowledge of the basis of production; on the other, a model of intellectual collaborative autonomy could be acted out in forums such as the Academy of Physics (1797). The practice of natural science in these contexts manifested a distinct if overlapping discourse, emphasising the purity of the academic commonwealth, transcending interested and privileged argument.[66] It is feasible to explore the further hypothesis that the concentration of republicanism in the arena of natural philosophy constituted an important means of political stylisation, orienting Edinburgh Whigs towards the Foxite opposition.

According to Forbes, Hume's scepticism was directed towards vulgar patriotic appropriations of constitutional history, and to the transient claims of political groupings. Scottish philosophical history, in close association with political economy, was not properly consonant with Foxite constitutionalism. In this respect, Leslie Mitchell's claim that nothing about Scottish ideas demanded modification of basic Foxite doctrine seems too sweeping.[67] As Donald Winch points out, the relation of political power to the progressive development of agricultural and commercial systems posed certain dangers to the Foxite view of politics. If commercial society could flourish under different political arrangements, and if, therefore, the process by which the English people had come to enjoy limited government could be interpreted as an accidental departure then whiggery's claim to legitimacy from the Revolution Settlement was vitiated.[68] However, leading figures in the party continued to emphasise the primacy of constitutional distinctions. Holland cited Rousseau approvingly to this effect in 1793.[69] The general election of 1784 produced a watershed in British politics, and there were plenty of grounds for supposing that the mantle of administering commercial society had devolved on the liberally minded William Pitt. In the wake of the Pittite landslide, Fox's followers faced a government that rhetorically embraced Smithian principles, negotiated the Eden trade treaty (1786–1787) and overhauled the system of taxation. Encumbered with the remnants of a country ideology, the Foxites had to make do with suspicion of the court, the reversionary interest, Burke's endless pursuit of Warren Hastings and sniping at the government's foreign policy.[70]

In this context it is useful to note that the Foxite alignment of John Millar (who dedicated his *Historical View of the English Government* to Charles Fox in 1787) was less inevitable among the Scottish intelligentsia than has been assumed.[71] After all, it was perfectly possible to extol the revolution of 1789, deplore that of 1793 and remain a liberal Pittite thereafter. We know from his correspondence that Francis Horner (1778–1817) plausibly defended his support for the government in the later 1790s. He wrote, 'Undoubtedly, within the last few years, violent attacks have been made upon the rights of the subject; but no-one finds his comforts impaired, nor his property less secure.' In 1799, Horner condemned Foxite opposition members who 'pant after a new republican order'.[72] Meanwhile, Henry Brougham flirted with Wilberforce, and the *Edinburgh Review* edited by Francis Jeffrey did not firmly commit itself to the Foxites until the eve of the formation of the Talents ministry (1806–1807).[73] Indeed, the northern intelligentsia were far from immune from a shift towards conservatism, as Sir James Mackintosh's partial retraction of his *Vindiciae Gallicae* indicates.[74] Personal relationships between the Whig aristocracy and their new Scottish professional allies could also be tense – there was more social equivalence between Henry Brougham and George Canning (1770–1827) than between Brougham and Lord Holland.

Yet the fact remains that the Foxite opposition did eventually claim the allegiance of the next generation of Scottish enlightenment intellectuals, lawyers and political economists. Most of the politicised Scots from the universities and the law did not become Canningites. Hatred of the Melville interest, shared opposition to the French revolutionary war, and the rhetoric of civil and religious liberty provided common ground. The war made the Foxites the sole champions of parliamentary reform besides the radicals, but there is little evidence that either the Association of the Friends of the People (1792) or Grey's reform motion of May 1797 was seized upon by the pupils of Dugald Stewart. So how was the Foxite alignment justified and displayed despite the competitive attractions of liberal Pittism? Science might help to throw some light on the problem, in an indirect way. Henry Brougham provides an example. His scientific writing shows how republicanism and science could be integrated to illustrate and argue for a political pose.

Brougham versus hypothesis

In April 1853, the Lord Chief Justice, John Campbell, embarked on the life of his intermittent legal and political rival, Henry, Lord Brougham. Looking back over a career spanning half a century, Campbell admitted that 'Volumes to load many camels might be filled with detailed accounts of all the doings, writings and speeches' of his subject.[75] Yet Campbell's aim was to show the ephemeral character of the former Lord Chancellor: 'He accomplished nothing as a statesman; he cannot be said to have extended the bounds of human knowledge by philosophical discovery; his writings, although displaying marvellous fertility, are already falling into neglect.'[76] Certainly, the protean range of Brougham's activities justified Campbell's likening of him to 'the Grecian Hercules, to whom the exaggerated exploits of many different individuals are ascribed'.[77]

It is suggestive that Campbell should have sought to diminish Brougham's contribution to 'philosophical discovery' – a backhanded tribute to the degree of his engagement with the natural sciences. So far as Brougham's initial struggle for political location is concerned, we are particularly interested in his early writings for the *Edinburgh Review*, from 1802 to the aftermath of the fall of the Talents Ministry, by which time Brougham's Whiggism had become consolidated (1807–1808). Brougham's scientific articles were produced just at the time when the Foxites were debating the proper means of political persuasion. We have already touched upon Lord Lansdowne's discomfort with newspapers and vulgar oratory as late as 1822, a suspicion echoed by Lord Grenville.[78] But Lord Holland recognised the importance of the expanding print media and tried to monitor and exploit it.[79] The launch of the *Edinburgh Review* marked a watershed in Whig polemic by the dissemination of printed text, particularly in its orientation towards the Foxite Whig party over the following five years. This periodical became a prominent vehicle by which natural-scientific polemic was spread. The *Edinburgh* reached a circulation of 13 000 copies by 1814, and eventually inspired the competition of the conservative *Quarterly Review* and later the Benthamite *Westminster*.

It is therefore useful to survey the political style of Brougham's natural scientific articles in this burgeoning publication, an area in his reviewing *oeuvre* that has been largely neglected. This is somewhat surprising, given the growing historiography of Whiggism and political thought. So far as the *Edinburgh Review* itself is concerned, physical science is circumvented in Clive's foundational study,[80] while more recent work tends to subordinate natural philosophy to the theory of commercial society and the enlightened character of political economy.[81] Fontana's otherwise impressive study of the *Edinburgh Review* contains only one direct reference to the chemical, astronomical and optical articles that comprised a great deal of its output.[82] This lacuna has been obscured by the fact that most aspects of 1790s Edinburgh have been thoroughly examined, including Brougham's involvement in the University's Speculative Society and the Academy of Physics.[83] Yet following Campbell, modern political historians have had difficulty in integrating Brougham's texts into his politics. The early articles have been dismissed because their conclusions were often wrong and their tone vituperative, while little interest has been shown by political historians in the structure of Brougham's arguments. Echoing Arthur Aspinall, Chester New labelled Brougham's optical theories an *ad hominem* assault on Thomas Young.[84] In revising this line of interpretation, historians of science such as G.N. Cantor have asserted that, on the contrary, Brougham's natural philosophy as expounded in the *Edinburgh Review* was expressive of a particularly rigid interpretation of the Reidian school.[85] But in Cantor's examination there is little sense of the aesthetic and political resonance of Brougham's output. As a consequence, Cantor tends to overemphasise one strand of Brougham's method, focusing on the Young controversy.[86]

Brougham's interest in scientific matters was prodigious. His articles included a survey of *The Elements of Optics* by James Wood, three pieces attacking Thomas Young and the undulation theory of light, and a critique of William Herschel's account of two asteroids.[87] He reviewed miscellaneous physical subjects in the *Transactions of the Royal Society of Edinburgh*. He also applauded Joseph Black's Life and chemical lectures, arranged by Professor John Robison, and this text was to reappear in his later *Lives of the Philosophers* (1855). Brougham commented extensively on contemporary theories of heat, and tackled the vexed question of meteors. The latter review was amended slightly and included in the *Tracts Mathematical and Physical* (1860). While the number of such articles tailed off after 1807, it is important to place them in a longer continuum of work. During the 1790s, Brougham had read and published three pieces of scientific juvenilia under the auspices of the Royal Society of London. These comprised two papers in 1796 and 1797 on extensions to the Newtonian corpuscular theory of light, and one paper in 1798 on geometry. In 1808 and 1809, Brougham still found time to examine Humphrey Davy's Bakerian lectures on electricity for the *Edinburgh*, and in 1811 he went on to praise Davy's work on oxymuriatic acids and gas, and continued to pen articles intermittently over the next two decades.[88]

Accounts of natural-scientific practice in the early nineteenth century have sometimes looked forward to the intellectual revolution consummated by William Whewell and John Herschel in the 1830s: the forging, that is, of the hypothetico-experimental method.[89] It is important to bear in mind, however, that at the turn of the century, hypothesis and theory were often considered to be reprehensible (though this was more the case in Scotland than in England, where a style of mechanical speculation subsisted, after the manner of David Hartley). For example, in 1795 Dr Alexander Gordon of Aberdeen, the hero of puerperal fever, declared that he discriminated between opinions on the grounds of practice rather than of unreliable theory.[90] Such enlightened devotion to the certainty of nature was characteristic of the period and was by no means confined to medicine. Professor John Playfair warned that in all branches of Natural History, 'theoretical language should, as much as possible be avoided'.[91] Cantor connects this suspicion of hypothesis to Thomas Reid's common-sense philosophy, which was partly a response to Humean scepticism, and was partly derived from the Baconian rejection of *a priori* hypotheses that followed the analogy of the human mind rather than the analogy of nature.[92]

Cantor states quite rightly that in many of the *Edinburgh Review* articles, Brougham remained steadfast in an anti-hypothetical line of attack,[93] and ostentatiously assumed a Baconian mantle, articulated by the Newtonian slogan *Hypotheses non fingo* ('I do not frame hypotheses'). Cantor further asserts the salience of this Reidian interpretation of Newton to Brougham's criticism of Thomas Young. In the Bakerian lecture of November 1801, Young postulated that light was the undulation of a universal ethereal medium. Brougham objected that 'the theory of this new medium ... is [only] a change of name: it teaches no truth, reconciles no contradictions, arranges no anomalous facts'.[94] It was a mere hypothesis, 'a wild phantom of the imagination'. Brougham insisted that Newton's conjectures about ether in the *Optics* should be firmly separated from his 'strict inductions', and further that Newton himself had never thought of ether as an established datum. According to Cantor, Brougham stood in a Scottish methodological tradition – upheld by David Brewster, John Robison, Thomas Brown and John Playfair – that viewed with suspicion all unobservable ethers such as, for example, the medullary (or nervous), gravitational and luminiferous ethereal fluids.[95]

Yet as Cantor himself has argued, and as Rashid has corroborated, the Scottish tradition was changing by 1800, with the emphasis shifting towards the acceptability of hypotheses in the formation of valid generalisations. This trend was foreshadowed as early as 1788 with Robison's acceptance that either the wave or the corpuscular hypothesis of light might be granted a high degree of probability, so long as resulting predictions were confirmed by a large quantity of data. However, as Salim Rashid points out, it was Dugald Stewart, the intellectual mentor of the Edinburgh Whigs, who really began the process of confronting the traditional deprecation of hypothetical methods during the 1790s.[96] In his wide-ranging lectures on political economy, in 1793 and 1798, Stewart condemned anti-hypothetical zealotry on notional Baconian grounds, arguing instead that modest conjectures anticipated principles.[97] He affirmed that hypotheses permitted the formation of problems, and thereby facilitated the useful arrangement of observation and experiment. This contradicted Reid's most frequently enunciated position, that a hypothesis could be accepted if (and only if) it manifested the application of an observed general principle in a new situation. It is worth adding that Stewart advanced both sorts of pro-hypothetical argument in the *Elements of the Philosophy of the Human Mind* (1814). In view of these canonical developments, it is necessary to ask why Brougham should have insisted so rigorously on the anti-hypothetical position. This was the case not only in his critiques of Young, but also in those of William Hyde Woolaston, William Herschel and Count Rumford. Cantor accuses Brougham of having failed to appreciate the utility of hypotheses in generalisation and of taking an obsolete Reidian position, rather than adopting a more contemporary pro-hypothetical one.[98]

In fact Brougham was sometimes prepared to veer at least to the edge of hypothesis, on occasions that Cantor does not consider. In his article on the origin of meteorites, in January 1804, Brougham attempted to substitute his own analysis for the work he was supposed to be reviewing.[99] Dismissing Joseph Izarn's *Des Pierres tombées du Ciel* as 'a mere transcript of the documents which contain the facts of the case', he proposed to compare two classes of data: the first comprising testimony of persons who had come into contact with the fallen stones, and the second derived from a study of the stones themselves and their surrounding matter. If both inductions presented comparable anomalies, equally irreducible to the existing arrangement of facts, 'then we may rest assured that a discovery has been made – and the two methods of demonstration will be reciprocally confirmed'. Brougham was aware that this sort of anomalous discovery could only lead to negative conclusions, in this case that the stones could not have been thrown off by terrestrial volcanoes, nor formed in the upper atmosphere (Izarn's theory). However, 'an hypothesis may perhaps suggest itself, unencumbered by any of the foregoing difficulties'. His idea was that the stones might have been propelled by lunar volcanoes. Brougham assured his readers that 'we may easily imagine such cause of motion to exist in the moon, as well as in the earth'. Acknowledging the difficulty of conceiving how the rarity of the earth's atmosphere could produce the high ignition of meteoric bodies, Brougham was content to leave the idea as the most probable hypothesis, rather than present it as a proven case. 'If, however, a more extensive collection of accurate observations, and a greater variety of specimens, shall enable us to reconcile the discrepancy... a knowledge of the internal structure of the moon may be the splendid reward of our investigations.'

Now, this argument has a number of features that should serve to diversify our view of Brougham's inductive method. The first point to bear in mind is that Brougham firmly adheres to the delineation between unobserved and unobservable entities. This was central to the Scottish distinction between hypotheses as generalisations (acceptable) and hypotheses as speculations (definitely not acceptable). It was permissible to imagine the action of volcanoes and heat at depth on the moon because similar action was to be seen on the earth. Ethers, on the other hand, could only be posited on the basis of secondary phenomena. In this respect, Brougham conformed to an idea of generalisation that was shared by Reid and by slightly later Scottish thinkers such as Stewart. But the use of this generalisation, in its provisional state, to anticipate the course of future knowledge was closer to the positive (albeit moderate) hypothetical practice that Stewart had begun to urge in his revision of inductive method. Brougham enters, as it were, a grey area in which certainly the liberal, and perhaps the restrictive, forms of inductivism would justify his strategy.

Here it is just as well to remind ourselves of the conditions under which Brougham actually wrote. Three letters to James Loch, in August and November 1802 and January 1803, convey a picture of an author preoccupied with his grand project, an *Inquiry into the Colonial Policy of the European Powers*, and for whom the *Edinburgh Review* articles were an exasperating diversion.[100] Now these sentiments were clearly subject to humorous exaggeration. Brougham was pleased to mention to Loch that the Review had reached its second edition, and that he was labouring like 'the horse of knowledge' to keep up with his book, his articles and his legal activity. Still, Brougham's complaint on 28 January 1803 is characteristic in tone. The 'd–d, blasted ... brutal Review' had halted work on colonial policy for a whole month, to enable him to turn out a hundred pages of print, of which only seventy were actually used. These comprised the chief articles on 'all the Mathematics (except one), all the Chemistry or Physics, and a long one on the Balance of Power'.[101] Brougham was working at speed, under pressure, with the aim of producing vivid, impressive journalism. He was also delicately navigating the first stages of his public life, in the manner fully described by Robert Stewart, not yet with the Foxites but gravitating towards them.[102]

This then prompts the question, what are the distinguishing features of those essays whose anti-hypothetical agenda is most pronounced? It is plain from reading these pieces that the Reidian invocation of 'the strict and chastened system of inductive logic' was congruent with a powerful laudatory or negative republican style.[103] Where Brougham builds up his own argument, induction is left to do its own work. In contrast, in the assault on Young published in the *Edinburgh Review* of January 1803, Brougham stated that he wished to raise his 'feeble voice against innovations', and to 'recall philosophers to the strict and severe methods of investigation pointed out by the transcendant [sic] talents of those illustrious men [Bacon and Newton]'.[104] In contrast, hypothesis generally was 'a work of fancy'. As the value of a hypothesis depended on its agreement with every fact that can occur, either it was a description of all the facts, in which case it was redundant, or it required constant adaptation to the phenomena that it sought to explain. This 'perpetual fluctuation' was 'the common lot of theorists',[105] and Brougham claimed that Young had altered his analysis between his lecture of November 1801 and a subsequent paper read to the Royal Society in July 1802. Inconstancy was the sign of a 'fickle and vibratory mind':

> We demand, if the world of science, which Newton once illuminated, is to be made as changeable in its modes, as the world of taste, which is directed by the nod of a silly woman or a pampered fop? Has the Royal Society degraded its publications into bulletins of new and fashionable theories for the ladies who attend the Royal Institution?[106]

This last reference was a personal hit against Young, who had taken the chair in natural philosophy at the Royal Institution in July 1801 with the responsibility of delivering public lectures, to which women were admitted. But the force of the masculinist jibe was coherent with anti-hypothetical rhetoric in a wider sense. Theory sprang from 'a warm and misguided imagination', and was tantamount to a 'sham fight'.[107] It showed neither 'vigorous habits of attention', nor 'powers of abstracting and comparing'. In fact, hypothesis was 'the unmanly and unfruitful pleasure of a boyish and prurient imagination, or the gratification of a corrupted and depraved appetite'.[108] Finally, Brougham condemned Young's appeal to the names of Euler and Newton, proclaiming that 'We hold the highest authority to be of no weight whatever in the court of reason.' Lacking a secure inductive basis, hypothesis could be enforced only by a kind of effete intellectual coercion. It was the work of immaturity and femininity, a decadence of knowledge.

Brougham's stand against fashion was also an ostentatious pose of cognitive republican virtue. It was 'no small proof of Mr Davy's natural talents and strength of mind, that they have escaped unimpaired from the enervating influence of the Royal Institution; and indeed grown prodigiously in that thick medium of fashionable philosophy'.[109] Naturally, his 'strict and patient induction', and modesty in his theoretical conclusions, had borne fruit regardless of his title.[110] Brougham attributed to Davy the 'free and manly' political sentiments that would enable him to resist the pressures of a 'courtly age' to rename potassium and sodium in honour of the Church and sovereign.[111] Davy's actual political opinions were of course irrelevant. What was important was that he could be invoked in a public performance. Brougham was later to criticise Davy for being 'always, from his first essays ... a dillettanti [*sic*] in nomenclature', that is to say, for lacking the very temperance in theory that the author had previously garlanded him with.[112]

This polemical context of Brougham's anti-hypothetical argument has been missed by Cantor, and the composition of such rhetoric has been ignored by political historians, who have merely noted its vehemence. Another sense of the republican provenance of Brougham's diction is to be found in his mauling of Count Rumford in two reviews of July 1804.[113] Discoursing on Rumford's *Inquiry Concerning the Nature of Heat*, there is the familiar distaste for 'a paper filled with theoretical matter, abounding in pulses, vibrations, internal motions, and ethereal fluids'. But the critique intensifies in response to Rumford's *Account of a Curious Phenomenon Observed in the Glaciers of Chamounay: Together with Some Occasional Observations Concerning the Propagation of Heat in Fluids*. Rumford had taken the opportunity in this second piece to retaliate against critics who had doubted his theory of ascending and descending currents. Brougham observed that 'there is an aristocracy, as it were, in the Count's manner of treating his adversaries, which we do not very well understand, and which is little suited to the republican constitution of the scientific world'.[114] Rumford had challenged his opponents to give a better explanation for the phenomena that he had observed. In response, Brougham reminded the Count that 'there is no demand in science for hypotheses, as there is for commodities in a market, where, of course, if what is good cannot be had, what can be had must be taken'. Neither Newton nor Joseph Black had reasoned like this, or had used 'authoritative menaces'. Consequently, 'the imperious postulates' of the Count should be resisted.[115]

The value that Brougham placed on a masculine science demarcated from commerce, defended from arbitrary authority and preserved from the hypothetical intrusion of female and puerile imaginations indicates his oppositionist republicanism. They are the signals that we would expect to find, if Brougham's republican constitution of science drew upon the rhetoric of a free state; if, in other words, scientific enquiry followed the neo-roman model of authority extensively described by Quentin Skinner.[116] This very naturally leads us to ask, what sort of republic did Brougham envisage, and what sort of republicanism was he in a position to deploy? As it was a constitution that Brougham took for granted, and therefore did not spell out his thoughts, it is necessary to make an assessment of probabilities on the basis of circumstantial evidence. It seems fairly clear that the model was not Rouseauan, since the dominant value was independence rather than the formation of a general will. Cantor suggests that the *Académie des Sciences* at Paris served as an exemplar for the Academy of Physics, the scientific society that Brougham helped to set up at Edinburgh in 1796.[117] However, this point is disputed by L.S. Jacyna, who points to mid-eighteenth-century Scottish civic republican templates.[118]

Further circumstantial evidence can be adduced to support Jacyna's supposition. We know from a letter of 1792 to the son of the historian William Robertson (Brougham's great-uncle) that Brougham had read the first five books of Livy by the age of fourteen.[119] These were the texts recounting the emancipation of Rome from the Tarquin kings and the establishment of a free state in which there was equality under an *imperium* of laws. As Jacyna argues, in 1796 Brougham coupled his suggestion to Francis Horner that an Academy of Physics be formed with an imprecation against the objectionable aristocracy of the Royal Societies.[120] On the same subject, and in the same year, Brougham told James Reddie that equality was the essential badge of philosophical politics.[121] Meanwhile, throughout the later 1790s, Brougham and his university contemporaries were often ardent participants in the volunteer corps. This was not only a matter for the students. Brougham later reminisced about the time both he and Professor Playfair served together in the artillery volunteers.[122] So when sixty years later, speaking before the Social Science Association, Brougham praised the patriotism and zeal for liberty shown by the volunteer movement of 1859, he was speaking from a background of personal engagement with civic humanist modes of acting.[123] Anti-hypothetical invective could be drawn into this material, and deployed journalistically in a public republican polemical style.

Yet if the raw material was neo-roman, it was a model that was pliable and adaptable. In 1808, Brougham defended Humphrey Davy against the charge that he had accepted honours from Napoleon, stating that 'we have always kept in the view of our readers, that the commonwealth of science is of no party, and of no nation. It is a pure republic; and it is always at peace.'[124] Brougham wanted to have it both ways: access to a violently partisan rhetoric *and* access to a mode of lofty transcendence. Having claimed Davy for independence from the court, he then proceeded to find anti-revolutionary attacks on him outrageous. There could be no 'cries of faction', only a common pursuit of truth:

> and they who would break the equality, or disturb the tranquillity of those sacred haunts, deserve to be chased out of civilized society,

> as aiming at the destruction of the only pure, dignified, innocent feature – the only remnant of the Divine origin – which bad passions have left in the character and conduct of men.[125]

This 'pure republic' was utopian in that it had no physical location, and was therefore incommensurable with conventional patriotism. The 'pure republic' was instantiated in various forums (journals, societies), but it was not assimilable to any one manifestation. It was pacific, not martial; its virtue was manifested in the independent and disinterested manner in which knowledge was sought. Furthermore, Brougham's vision of the accumulation of scientific knowledge was progressive, not static; the practice of serene contemplation was juxtaposed with the material benefits that science was to bring. This, however, was set at a remove, for the commonwealth of science notionally existed in a state of innocence. Its moral significance lay in the character of men who inhabited it; such men as the chemist whom Brougham revered, Joseph Black, a person of 'calm and immoveable judgement'.[126]

The notional abstraction of the commonwealth of science from real politics is brought into focus in Brougham's mitigating plea for Lavoisier.[127] Morrell concentrates on Robison's blanket attack upon, and Playfair's vigorous defence of, the French innovations in nomenclature and metrology.[128] The point to add is that Brougham was here manipulating a republican discourse of real and sensible meanings masked by insubstantial and deceptive language, which by its nature had to be imposed arbitrarily by authority. The imposition of mensural language by *fiat* might be seen as absurd, but it was also harmless, whereas the imposition of political structures was highly dangerous. The Whiggish implication was that the source of arbitrary political imposition in Britain was as likely to come from the court as from revolutionaries. Robison had criticised Lavoisier's replacement of de Stahl's phlogiston theory, alleging that the change in nomenclature had been part of the same frenzy and inanity that had marked the introduction of the French revolutionary calendar. But Brougham preferred merely to ridicule the pedantry of French radicals, who failed to see the difference between the two cases. It was vain to think that altering ordinary language could transform society. However, the objections to abrupt political changes could not be brought against philosophical innovation. The new chemical nomenclature was implemented after chemistry itself had been transformed, whereas the new calendar was simply the product of revolutionary caprice. In this way Brougham preserved scientific from political radicalism. He poked fun at a dubious anecdote recounted by

Robison, in which Madame Lavoisier, in priestly garb, burnt Stahl's treatise on an altar, while solemn music played a requiem to the departed system.[129]

So, on the one hand, Brougham was ready to affirm the essential distinctiveness of scientific language and the need for it to mirror inductive truth regardless of tradition. On the other hand, this same language was applied to a separate realm. Its descriptive scale could be used to excoriate any offending aesthetic that was unnecessarily esoteric. In October 1804, Brougham savaged Young for his 'undulatory and vibratory mode of writing'.[130] Here again, the key was suspicion of the complicity of imagination and the passions. This was a connection that in a very few cases (Pascal and Newton) might produce genius, but which more often corrupted the scientific academy. It was Black's 'want of the... passions' in early manhood that gave him 'the most essential characteristics of the inductive philosopher'.[131] These included simplicity in description, thereby opening a given field of enquiry to 'universal scrutiny'. The same Stoic requirement was enjoined in a slightly different way, on Herschel's identification of the asteroids Ceres and Pallas. Brougham accused Herschel of employing 'compound epithet and metaphor', and condemned the astronomer's 'telescopic sweeps', his 'natural history of the heavens' and especially his use of the image of 'construction' to describe the fixed stars.[132] The last, Brougham thought, would encourage the vulgar notion of the heavens being a blue vault. What was at fault was 'the great power of words in misleading and perplexing our ideas'. This was the mechanism by which hypotheses exploited the imaginations of the weak, feminine and puerile. It was this that gave rise to the importance of 'plain doctrines... taught in the plainest manner'. This was one legacy of Bacon that Dugald Stewart brought home to his pupils such as Brougham: to distrust the Idol of the Market-Place, or the human predisposition to accept false appearances imposed by words.[133]

To sum up, Brougham's methodology was not quite as pure and rigid as has been described by Cantor. Brougham was prepared to utilise moderate hypothetical arguments about observable phenomena when it suited him; he was also more than ready to indulge in anti-hypothetical invective. Drawing on the Reidian canon, this rhetoric was also a set of republican political signals – in a flexible adaptation of the civic republican mould identified by Jacyna, and also in a transcendent, Stoic register. Brougham's scientific rhetoric thereby comprised an amalgam of elements. A common feature of this style was emphasis on the control of the passions and the speculative imagination. The style was also selectively arcadian and egalitarian, sometimes carrying echoes of Livy in its criticism of arbitrary authority. In this respect, we might look to Brougham's scientific articles in the *Edinburgh Review* to anticipate his more famous outbursts against monarchy and the nobility, such as his praise for popular revolution in the 'Don Cevallos' article of 1808.[134] As such, Brougham's scientific output served to align him with the commonwealth and oppositionist style in Foxite politics, but in an indirect manner that did not commit him, at such a fluid stage in his political career, to the Foxite cause. More fundamentally, Brougham's scientific polemic allowed him to uphold the concerns of Baconian sociability: the equality of philosophic discourse against aristocratic pretensions, the stability and chastity of language and of course the wider publicity of improving knowledge.

3
Rational Sociability

Foxite materialism and Holland House

Whiggery can be cast as progressive or conservative according to its attitude towards established interests and institutions. Similarly, it can be defined as liberal or non-liberal according to its relationship with public authority and public finance. To see yet another dimension of political definition we have to consider Whig attempts to signal what might be called their rational sociability (a concept which gained politico-ideological force from the development of the idea of the statesman). We can access this aspect by adducing the sociable frameworks within which Whigs approached cooperation with each other and other groups.

Variants of rational sociability can be demonstrated by comparing two prime locations of aristocratic influence: Holland House and the Lansdowne estate at Bowood. In the metropolitan milieu of Holland House, the principal Foxite salon between 1797 and the later 1830s, there was a close correlation between the perceived constitutional definition of Foxite identity on the one hand and literary, library-bound science on the other. The static obedience that supporters of the court were portrayed as showing towards the monarch was the antithesis of Foxite rational debate, according to Foxite self-perception. Thomas Macaulay eulogised Lord Holland by lauding his courtesy, pleasant disputatiousness and independence of mind. A complaisant House of Lords compared unfavourably in a discursive exercise to Holland's own circle, which brought together the talents, arts and sciences in conversation.[1] Natural science was of interest to the extent that it served a reasonable sociability and a wider aesthetic connection.

Both Leslie Mitchell and L.S. Jacyna have noted that speculative physiology and psychology had an entrenched presence at Holland House in the person of John Allen (1771–1843), salon librarian between 1802 and 1843.[2] Allen generally promoted the materialist doctrines of Hartley, La Mettrie and Holbach. His beliefs were gratifyingly dismissive of final causes and complemented Holland House scepticism towards organised religion. Enquiry over the nature of bodies formed little part of Charles Fox's own purview, however, so it is worth asking where it was that Lord Holland got his taste for materialist mechanics in the first place, especially as this is a topic that Leslie Mitchell passes over.[3] There is evidence to show that by late 1800 Holland was familiar with William Roscoe's circle of Liverpool literati, a group which included the Edinburgh-educated Dr James Currie (1756–1805).[4] Currie continued to espouse iatrophysical modes of explanation, even as this quantitative and mensural approach was decisively falling out of fashion in the later 1790s.[5] In 1807 Holland asked Roscoe for details of local Liverpool reading societies.[6] However, the strongest indications point to the influence of Holland's Hispano-Italian tour (1793–1796), specifically the two years that he spent in Florence from the autumn of 1794 in the company of Lord Wycombe (1765–1809), Shelburne's eldest son and elder half-brother of Henry Petty.[7] Holland admitted that he never saw 'Spallanziani' (probably the natural historian Lazzaro Spallanzani, 1729–1799) nor enjoyed the acquaintance of Galvani (1737–1798), whom he met briefly at Bologna,

> but of Fontana, whose contemporary reputation was greater than the works he has left will maintain, I can speak from personal intercourse amounting almost to intimacy.[8]

Felice Fontana (1730–1805) had been summoned from his chair of physics at Pisa to Florence, in order to reorganise the grand-ducal collection of Leopold I (1747–1792).[9] From 1775 he was director of what Holland called the Natural History Museum (the Museum of Physics and Natural History). Holland asserted that Fontana had frequented Holbach's salon, was a covert Encyclopaedist and was a materialist 'after Helvetius'. He belonged, Holland wrote, to a party of Tuscan freethinkers opposed to the Jesuit and monastic orders in Florence. It is certainly the case that Fontana was briefly imprisoned in 1799 for his French sympathies. Most significantly, Holland described the 'anatomical man' fashioned out of pieces of wood by Fontana's deputy Giovanni Fabroni (1752–1822), which Fontana demonstrated 'as if his own'.[10] The reference to a wooden body is suggestive, because Fontana was also widely known for the production of wax anatomical models under his directorship.[11]

Now, these memoirs were written after 1826 and need to be interpreted. Holland's reminiscence was intended as an exercise in period portraiture, self-consciously situating his own person at the centre of contemporary style: 'natural philosophy, however, was the passion of the day, and geology, chemistry and electricity the fashionable topics of conversation'.[12] The portrait was structured so as to blend familiarity and distance. It was therefore important for Holland to show that he was conversant with Godwin's scheme for the perfectibility of man, and Erasmus Darwin's *Zoonomia,* at the same time as emphasising his lack of intellectual fealty to those, including the Liverpool circle, who were 'stigmatised as visionary speculatists or disaffected Jacobins'.[13] This was the context of his depiction of his 'intimacy' with the Italian savants – 'I was myself far too ignorant to judge of the professors or proficients in such studies.'[14] The argument is not, then, that Fontana advocated an iatromechanical programme (in itself contestable) that Holland then adhered to: as Goethe's appreciation of Fontana's waxworks illustrates, such modelling could be appropriated for distinctly un-whiggish intellectual projects.[15] Ideological transmission was not the most important part of the relationship. Instead the key technique of Holland's stylisation was locating his position by showing knowledge *of* such trends and the figures that represented them. In this, there is a clear parallel between Holland's dilettantish relationship with savants and the relationship of the third Marquess of Lansdowne with the canonical texts recommended to Stanley, noticed previously. We are seeing the difference between using knowledge in order to form or buttress ideological purposes and the *allusion* to knowledge, even knowledge not properly or professionally understood, as a badge of entitlement to statesmanship.

Evidence from Holland's correspondence with Shelburne shows an acute awareness of the possibilities of political positioning by natural philosophic doctrine. There are no extant references in the Holland Papers to Fontana or Fabroni but in August 1794 Holland wrote that one Andreani was most kind to him at Milan despite his busy activity in building observatories and pursuing natural philosophy.[16] Andreani was reportedly finding that political pressures were closing in on him amid the Milanese apprehension of revolution, which reflected the poor character of Milanese government and its vulnerability to open scrutiny.[17] Lansdowne replied that he respected Holland's friend and that he was glad that he had taken to philosophy politically, noting that Benjamin Franklin had used philosophy to clothe his political views.[18] The exchange suggests that the friend in question was the

Cavalier, or Count Paul Andreani (1763–1823), the Milanese balloonist, mineralogist, North American explorer and republican sympathiser. Andreani had returned to Milan in the autumn of 1793, after the latest of his repeated sojourns in London (1784, 1789–1790, 1792–1793).[19] It is possible that Shelburne had met Andreani at the dining club of the Royal Society, a milieu frequented a little earlier by Franklin.[20] Whether through the Italian examples that he encountered or Lansdowne's illustration of the nexus, Holland was educated in the political signals of natural philosophy. This is the background against which his later patronage of Allen, as explored by Jacyna, should be set.

There were two aspects of scientific style: positioning by association with savants (which was consistent with a dilettantish approach similar to Holland's) and intensive public or published exploration of ideas (of the kind entered into by Brougham, for example). It was not necessary for statesmen to be fully conversant with the details of the positions that they patronised in order for them to send a series of political signals through their association. The point might seem elementary, but it is worth pressing, because it helps to explain the intellectual connections of political actors who were not necessarily deeply versed in the theory or details of their cultivated subjects.

Foxite materialism is not to be understood as a fully worked-out doctrine in the hands of its patrons. It was a style, which in Holland House hinted at anti-clericalism, liberality and subversion without presenting a doctrinaire critique. Of course, as Jacyna shows in the case of Allen, it was certainly possible for a unifying thread of ideology to run through the thought of protégés.[21] On the other hand, there were central themes which were relatively easy to pick up and deploy. One such theme was the self-sufficiency of nature. A corollary noted by Brent (though not with its corporeal context) was vehement rejection of the Trinity, which Holland described as 'revolting to his understanding'.[22] Not having a very strong commitment to divine superintendence beyond creation, unitarianism could be embraced as a distanced, provisional theology of rationality and social stability.

Holland House was not, in any case, a place of physical experiment or base for outdoors investigation. The rising young generation of Edinburgh Whigs entertained there were not celebrated as practising geologists, despite their considerable interest in the subject. In a sedentary, artistic and literary circle, in which the country was held to signify ignorance and squirearchical crudity, the pursuit of observation out of doors was deprecated.[23] Authors notable in science, art and poetry might merit audience and discussion, but Holland House stood apart from the muscular exercise needed to go over rock formations and soils. Under the aegis of Lady Holland, female sophistication blended the library with the drawing room.[24] The leading actors were men, but they were held to inhabit a feminised interior space.

Leslie Mitchell has described the disdain felt by Holland House Foxites and Edinburgh Reviewers for the Lakeland poets, their supposed idolatry of nature and attribution of elevated sentiments to social inferiors.[25] This stood in contrast to the impulse-driven Romantic geology of geologist-liberals such as George Bellas Greenough, MP (1778–1855), the first President of the Geological Society of London.[26] Greenough was favourable to Catholic relief and sinecure reform, but stood apart from Holland House and was wooed by the Canningites.[27] If Geology was, as John Wyatt argues, a poetic discipline, it belonged to an aesthetic that Foxites found uncomfortable.[28] It was associated with capricious, individualistic and uncivil politics. The nature poet whom Lord Holland favoured was the neo-Augustan George Crabbe, rather than Wordsworth, Byron or Coleridge.[29]

Geology was also suspect for another set of reasons which were manifested rather earlier during Holland's European travels. Studying the fertility and properties of the land was seen as a tedious department of agricultural and extractive political economy. Holland complained to Shelburne in September 1793 that the otherwise informative *Journey Through Spain* (1791) by the Reverend Joseph Townsend (1739–1816) was made dull and uninteresting through being stuffed with geological rubbish.[30] Townsend's work promised 'particular attention' to the 'agriculture, manufactures, commerce... and revenues of that country'.[31] The survey presented the 'Natural and Civil History' of each region according to its stage in Townsend's Spanish itinerary.[32] Shelburne encouraged the young Holland to use his Continental tour to extend his acquaintance with practical matters of estate management and political economy. Holland was advised to master the Spanish wool trade and the disruptive effect of speculation on prices.[33] Shelburne wrote to Holland that there was an urgent economic imperative for such a study, as both men's estates were likely to suffer due to war and poor harvests.[34] Interestingly, Holland was also told by Shelburne to instruct himself in the management of cultivation and trade in agrarian commodities for the public weal as well as the interests of his own property.[35] That is, Holland's instruction was conceived in terms of his nascent role as a public statesman, as well as a private landowner. It was in this context that Townsend's text was introduced. Holland's intellectual priorities were those of his more influential uncle and mentor,

Charles James Fox. They were overwhelmingly classical and historical, and certainly did not include the earth sciences, chemistry or economics.[36]

The Lansdownes

Holland's early correspondence with the Earl of Shelburne furnishes an initial point of comparison with the rival seat of Whig patronage that developed at Bowood in Wiltshire. Bowood was a centre of experiment long before Shelburne allied with Fox in the 1790s. Joseph Priestley (1733–1804) was librarian there from 1773 until 1780, and pursued an active scheme of laboratory research at nearby Calne. 'Lord Shelburne encouraged me in my philosophical inquiries, and allowed me 40*l* per annum for expenses of that kind, and was pleased to see me make experiments to entertain his guests.'[37] Unlike the London setting of Holland House, Bowood was physically situated in proximity to an important cluster of provincial geology. Townsend was within reach as Rector of Pewsey, and was connected to Shelburne through Jeremy Bentham (1748–1832).[38] Under the third Marquess, geological excursions became a Bowood speciality. Charles Lyell and Leonard Horner were both habitual visitors, and the former was ready to give Whig ministers guided tours in the vicinity 'from the oolite to the chalk downs'.[39] In this respect, the Lansdowne circle approximated much more closely than Holland House to the received picture of masculine nineteenth-century 'gentlemanly' science. Writing to Lord Auckland in 1816 of his trip to Vesuvius in the company of John Playfair, Lansdowne noted that the site had generated some copper, which was a volcanic novelty.[40] This was a physically engaged active science. Outwardly oriented activity was pursued across a wider front. Praising the new breakwater at Plymouth as a most satisfactory public construction, Lansdowne told Auckland that he had spent a day on it with the engineer, who was happily and successfully laborious.[41]

Lansdowne's association of purposeful activity with administrative vigour had a different atmosphere to Foxite discussion. The action of the Dey at Algiers, Lansdowne confided to Auckland, in personally officiating at the repair of the docks, showed that he had some of the great as well as bad qualities of a tyrant.[42] So far as British politics was concerned, parliamentary debate was unlikely to deliver what even confidently liberal government could not. The nation was to be extricated from its difficulties not by any one measure but by steady and prudent administration.[43] Lansdowne fostered a style of technical, liberal politics that stood somewhat outside the Foxite mainstream. In part he transmitted the legacy of his father, who had moved closer to rationalist aristocratic radicalism than to Foxite 'country' archaism during the 1790s. Shelburne had dallied with men such as Stanhope and Lauderdale. When Stanhope wrote to Shelburne in November 1794 to profess his public zeal, he did not mean a return to Miltonic virtue. Instead he expressed interest in the continually increasing 'ratio' by which public opinion concerning politics was changing.[44] Shelburne had combined a future-oriented temper of belief with a degree of realism, confessing to the Duke of Grafton, for example, that the public response to changing conditions fell far short of opposition hopes.[45]

The important thing to notice is the difference between scientific activity at the different locations of Bowood and Holland House. For the Lansdownes there was a fairly evident bifurcation: a vigorous, practical interest in matters such as engineering for the purposes of public administration and an extra-curricular leisured enjoyment of science, both in discussion and in physical activity (for example in geological jaunts). In Holland House, on the other hand, the salon environment gave every form of scientific discussion a politicised subtext. It was an aestheticised experience, and subordinate to artistic and more directly political topics. These styles of science expressed different kinds of Whig identity: one foregrounding executive action and the other finding its political formation in sociable exchange.

Now, if one regards the politics and science of these two locales as discrete entities, one is left with problems of explanation that are necessarily extremely difficult to determine. Because these entities are coterminous, it is impossible to say which came first. Instead, it is better to think of these phenomena as integrated styles, expressed at – and expressing – particular locations. The process can be seen in two of the letters cited above, composed within six months of one another.[46] When Lansdowne clambered about on Plymouth breakwater he was enacting the convention that public work testified to industry, and should be recreated by the man of affairs. By articulating this episode to his ally Auckland, he was signalling his adherence to this statesmanlike posture. This was not a lesson taken from engineering or the transfer of executive politics to building. It was political location by public style.

Knowledge and social leadership

Much has been written about the Broughamite educational organisations – the Mechanics' Institutes, the Society for the Diffusion of Useful

Knowledge (SDUK) and the London University.[47] What is missing from these accounts is any discussion of established Whig models for marshalling and leading opinion, the most crucial element of which was the voluntary deference shown by independent men. It has often been pointed out that the town or county meeting was the favourite milieu of Whig notables,[48] and the parliamentary petition (the visible gathering of names) the preferred mechanism of political advocacy. Undoubtedly, this was the oldest and simplest way of securing the consent of 'independent opinion' in a controlled manner, especially in those counties that were not readily susceptible to aristocratic manipulation. Such forms of demonstrative assembly conferred high status on public action. Correspondence between Joseph Jekyll and Shelburne attests to the role played by meetings of electors in Westminster, petitioning and resolutions in the Whig club in maintaining Fox's standing.[49] The resolutionary meeting was clearly important for the public status of the statesmen who participated. Moreover, the political historian is struck by certain structural similarities (as well as the obvious differences) between the form of the town or county meeting and the later scientific and educational societies. The main similarity lay in the link between opinion and independence. The ordinary attendees at the town or county meeting gave stature to the statesmen who interacted with them because they supposedly represented independent opinion – capable, for example, of being represented in a petition. By comparison, while it is true that (as Desmond argues) the social groups approached by the Broughamite Mechanics' Institutes and SDUK were subordinated,[50] nevertheless it was not the case that Brougham saw working men and other targets of the Institutes and SDUK as passive or empty vessels. What was important was that they should recognise, understand and defer to the knowledge that they received. It was, after all, the power and capacity for action of the target classes that gave educational initiative its social utility. The Coleridgean insult in the *Athenaeum* that the SDUK and London University were run on a "close borough system" was a jibe which gained its force from the Whig idealisation of the consent available from open constituencies and forums.[51] Shapin and Barnes are right to observe that Brougham's urging of majority worker representation on the Board of Directors of each Institute hardly amounted to 'workers' control', but they are mistaken to imply that this was a deceitful move.[52] On the contrary, the presence and active collaboration of the target group corresponded to the traditional hierarchical mobilisation of opinion, as well as its instruction and formation. The guiding problem was how to turn artisans into free citizens: to cultivate their independence and then to lead it, much as Whigs sought to do with the pre-Reform electorate.

Thus Brougham praised the managers of the Manchester Mechanics' Institution for operating schools where boys 'are taught to reflect and reason', and urged 'mutual intercourse and social study' as the path to emancipating workers from a mechanical existence, in their unreflecting and rote labour (July 1835).[53] Knowledge and saving were the means by which the artisan might 'make him[self] an independent man, without which he does not deserve to be called a man at all ... much less the citizen of a free state'.[54] On laying the foundation of the Liverpool Mechanics' Institute in the same month, Brougham delivered a paean to the educational underpinnings of civil and religious liberty,[55] which he believed were pacific and pastoral rather than martial and industrial. He specifically decried the phrase 'march of knowledge', preferring the horticultural image of the 'schoolmaster' tending the garden path of cognitive improvement,[56] and he insisted that the comprehensive spread of knowledge was the only guarantor of peace and trade. This was a quintessential expression of scientific liberality. Having just been left out of Melbourne's second cabinet in April, Brougham went on to treat the gathering to something like a stump speech. He 'had not found favour with the courtiers', but his educational work in the SDUK and the Mechanics' Institutes would enable him to resume his 'position in the van of [his] countrymen'.[57] Rhetoric, certainly, but his style had the effect of projecting the scientific organisations in such a way as echo a parliamentary constituency.

Coalitions of knowledge

The idea of coalition, and its resonance with underlying ideas of party, had reverberations across the field of scientific activity. Whiggery was in the first place a politically sociable, *party* identity. In a corporate sense, the ties at the heart of Whiggism, the genealogical and interested connections between 'Friends' and families, constituted political bodies. Edmund Burke described the Hanoverian Old Corps Whiggery as a tight-knit nexus of property, mutual obligation, blood, popular affection and dynastic loyalty.[58] It was both a body of power and a body of men promoting the national interest on the basis of agreed principle.[59] The formula indicated that while the party-coalition model could include dependents, it was fundamentally oriented towards the sociability of peers. This was in contrast to the vertical structures representing social subordinates discussed above. The present focus is not so much on the structure of Burkean thought as on the continued potency of such Rockinghamite assumptions for the early-nineteenth-century Whig party. Nevertheless, Burke's emphasis on connection as a mode of rational self-formation is important to note. Detached gentlemen might be angels or devils, but Burke informed his readers that party was a means by which he and they might be formed for public virtue. Embodiment was a contingent choice in response to a public need, and could be mistimed. Men might see the need for honest combination, but see it too late: they might form a body when it would prove ruinous to them.[60] The body constituted by party was an agent of ordered motion in a public life of power and energy to be contrasted with the rage of civil violence or despotic repose.[61]

Party was the expression of a dual ontology in which social matter was guided by agreed principles. This was the moral force of the union of public men, as Henry Grey Bennet put it, counter-posed to the arbitrary authority of the Crown.[62] Principles without the body of men to carry them through were nugatory, whereas a connection without principle would degenerate away from disinterest. Party was a form of organisation rather different to a market, as it was created by the collective assent of conscious volitions rather than being the spontaneous product of individual interactions. On the other hand, like the market, it validated inequalities through the egalitarianism of its principled structure. An excellent illustration of the operation of this social model (or rather the stylised way that this model was meant to work) is provided by Lord John Russell's sketch of a Tory politician under Queen Anne:

> He adopts, if you please, the general opinions of the tories. He votes generally, but not always with that party. He naturally becomes acquainted with some of them. He talks over the questions that are coming on for some time before. These conversations lead to a more intimate union: his opinions are listened to, and his doubts melt away in the course of amicable discussion. Sometimes, when the measure is one of party policy rather than of principle, he surrenders his own opinion to that of the statesmen most respected by the society of which he is a member ... He is, in short, a party man.[63]

As Austin Mitchell justly points out, such nostrums were really descriptive of the nineteenth-century world of party.[64] The structure allowed the development of deliberative or discursive authority. For example, it permitted Charles Grey to acknowledge assistance and direction from those he respected while preserving his independence.[65] The idea of a common adherence to principle was a passport to the formation of rational deference. The boundary between leading and material principles (which demanded universal adherence) and points of minor importance (where occasional concessions were possible) was, in practice, conveniently vague. What was crucial was the idea that such a demarcation existed and that its presence legitimated the evolution of deliberative ties. This idea simultaneously confirmed social inequalities and the autonomy (independence) of 'honourable men'.

This style was extremely flexible, and although the gap between principle and expediency (the sacrifice of reasoned adherence to unthinking appetite, usually for the goods of office) was a point of potential rhetorical weakness, the logic could be extended to cover cooperation with other public men who were not necessarily natural allies.[66] This was the kind of critical relationship that Whigs were eventually able to form with Peelites in the later 1840s. The rationale of coalitions could be subscribed to by statesmen who would not consider themselves party men in the narrower sense. As the past President of the London Geological Society Henry Grey Bennet put it, when speculating on a possible understanding between the Whigs and Canning in 1821, men could hold their principles and act together without agreeing in all opinions. Indeed, Bennet surmised that he differed as much from fellow oppositionists such as Lord George Cavendish as much as with Canning. Party itself was a spectrum that could be extended to a coalition.[67] This was the logic that Roderick Murchison was calling in aid when he appealed to Brougham to tolerate Oxford tories within the framework of the British Association, because it amounted to a coalition with them (1832).[68] Coalition was the outer circle of political sociability, implying concert without necessarily deference in the promulgation of views. By comparison, the inner unit of the political body made by Whiggery was encapsulated in the word 'Friend'. The Foxites were Friends of the People, Friends of Reform or Friends of Peace. As Friends, Whigs indulged in more than friendship and conviviality, though these were certainly prominent elements. Friendship expressed the sociability of party principle, integrating rational deference and independence.

In other words, this style of (inter-) partisan sociability was distinct in its ideal form from the dematerialised republic of letters that Lorraine Daston identifies as an essential step in the moral history of objective science.[69] The model of friends and coalitions was realised far more in the disposition of the Whig Liverpool Concentric Society (1812), with its ordered rings of freemen and notables,[70] than in the philosophical society identified by François Furet, in which members shed their social distinctions to enter a community of ideas.[71] The essence of Whiggery, on the other hand, was that participation in its ideal sociability was a matter of ideas (or principles) and concrete distinctions. When Whigs created or joined natural-scientific forums, they took this conception with them. This is not to say that when Whigs joined societies they necessarily sought to dominate them as a faction or party. For example, the group that joined the Geological Society of London in 1813 in no ostensible sense acted together as a clique among the other members.[72] Likewise, writing to Charles Babbage in the aftermath of the bitter battle over the election of the Duke of Sussex as President of the Royal Society in 1830, Henry Brougham attempted to restrain further polemical controversy.[73] When it suited, Whigs were perfectly prepared to proclaim loudly the transcendence of faction that scientific cooperation implied. In such cases the pursuit of the truth performed the same function as the principle of restraint of the crown, or Catholic Emancipation; it was the constitutive goal that (at the appropriate moment) legitimated coalition with men of very different persuasions in other areas of policy or endeavour. This transcending strategy was a political mode of coalition building, with its background in the heritage of arguments over party.

Coalitions for the truth required a standard of unity beyond the *ad hominem* deference paid to trusted eminence, and this was where the 'objective' calculus of facts entered the frame. The tribunal of the facts offered an excellent stand upon which to build coalitions, not because the identity of the investigators was abnegated, but because perceptible, communicable facts offered a provisional and shared sociable process. It was a canon of coalitionism that none of the parties need give up irreconcilable tenets that were nevertheless compatible with specific principled aims. Laying disagreement to one side, in the name of a provisional procedure, was how this sort of extended sociability functioned. The phenomenon can be illustrated through attention to the parliamentary membership of the London Geological Society, which was an example of an intellectual coalition in action.

Seeing the Geological Society as an example of intellectual coalition-making varies the way it has been previously analysed in terms of high politics. Rather than a culture of coalition-making, the structural analogy of Westminster has been applied instead: a model of proposition and opposition in educated conversation. For James Secord, the Geological Society was an example of club land culture dominated by the grandest of all clubs, Parliament.[74] Like other clubs, its social calendar was that of the London political season.[75] Secord's point is corroborated by Hume's observation that 'the period of the year during which almost all the societies hold their meetings coincides in a great degree with the parliamentary session'.[76] Geological discussion was exclusive and intimate. It was preserved from the press with an intensive orality noted by contemporaries. This oral communication network, as Secord puts it, was intrinsically political.[77] Both Secord and Martin Rudwick emphasise the importance of genteel oral codes in the early-nineteenth-century geological world, where at the same time parliament represented a regulative ideal of public life.[78] Paul Weindling compares Banks' relief in 1804 that the Royal Society was free of addiction to politics with Charles Francis Greville's jocular parliamentary label for George Bellas Greenough, first President of the Geological Society, and MP from 1807.[79] Nicolaas Rupke refers to a series of adversarial debates at the Society (on glaciation and the theory of the earth's central heat) to suggest that by 1840 the Geological Society had become a key arbiter of geological opinion.[80] Meanwhile, Rudwick also points out the character of the society as a corporation based on private subscription, compared to the state-backed Paris Museum.[81] Thus the society drew upon the same aristocratic and aspiring middle-class constituency that dominated the political elite.

One can certainly see the force of the parliamentary analogy, but there is something missing from an account that goes from drawing room or salon to club (a fair transition) and then jumps to parliament. The intermediate structures of party and coalition are the missing cultural links. The very intimacy of the Geological Society's conversational culture noted by Secord is more reminiscent of party negotiation and conclave than the formalised and large-scale debates of the Commons, which were at least partly oriented towards a wider political public 'out of doors'. The society disavowed sectarianism in its rhetoric, but then so did the Burkean party in its disinterested conjunction of independent men. The self-declared 'Parliament of Science' in the 1830s was the British Association, which was instituted on a broader scale. This Fourth Estate notionally included *representatives* of all the different scientific societies.[82] Morrell and Thackray describe how the gentlemen of the specialist societies sought to make use of the Association's forums of policy and patronage.[83] The Geological Society did indeed function within a parliamentary culture, but this culture was rather denser in its intermediate structures than the concept of parliament alone.

Perusal of the membership and officer lists of the London Geological Society discloses a roll of names also active in the House of

Commons.[84] The following parliamentary prosopography complements the thorough accounts of the structure of the Society that already exist.[85] Between the foundation of the Society in 1807 and its incorporation in 1825, twenty-four sitting MPs were elected to membership or honorary membership. There were eight men who might be described as consistent Whigs, one radical political economist (David Ricardo) and two who generally voted with the opposition on liberal measures. There were eight habitual government supporters, three semi-detached ministerialists and two who are difficult to classify, even within these broad terms. But this summary scarcely does justice to the fluidity of political positions espoused by this sample. Take one of the semi-detached ministerialists. William Jacob, MP (1762–1851) was a South American merchant who sat for the Treasury borough of Rye in the year of his election to the Geological Society (1808). Jacob invoked the chemistry of minerals during debates on the salt duties in 1809, and was classed as 'Govt.' in Whig division lists in 1810. He was essentially a late representative of the regime Whig tradition, who contested Great Yarmouth in the Duke of Portland's interest in April 1807, afterwards resisting sinecure and parliamentary reform. Yet Jacob also urged the independence of the Spanish American colonies, advising Fox on the question in 1806. He condemned the slave trade, was a moderate free-trader, and even claimed to be independent of the court.

Of the eight steady and three arm's-length government men, one was a commissioned army officer (Lieutenant-Colonel William Congreve), two were country gentlemen (Viscount Kirkwall and the distinctly unsteady Richard Vyvyan), one was a Norwich corporation patrician, wool merchant and banker (Charles Harvey), two had place-holding backgrounds (the Scottish Melvillite Sir George Clerk and the Chancellor of the Irish exchequer, John Foster), and three others besides Jacob were connected with oceanic commerce or colonial diplomacy: the East India diplomat Sir George Thomas Stanton, who sacrificed his close borough over Catholic Emancipation in 1826; Sir Charles Long, a City merchant dealing in the West Indies trade; and Charles Milner Ricketts, former East India Company writer and client of the Earls of Liverpool.

In contrast, the cluster of eight Whiggish and two liberally inclined Opposition members tilted decidedly more towards the land, that is towards both agriculture and mineral extraction. Francis Horner and Henry Grey Bennet were lawyers, while Joseph Foster Barham inherited Jamaican planting interests and a Pembrokeshire estate. John Frederick Campbell (member for Camarthen, 1813–1821) succeeded his father as Lord Cawdor in 1821, the year of his accession to the Society, and came to possess considerable estates in Nairnshire and Cardiganshire. While one of the sympathetic independents (Sir John William Lubbock) was a London merchant banker, the other was a known agricultural improver in Hertfordshire, Sir John Sanders Sebright. There were two country gentlemen: William Taylor, the MP for Wells, and the distinguished Scottish corresponding member, Sir James Hall. They were joined by two rather different landed scions: John Lambton, who inherited 17,000 acres in 1813 replete with coal; and Viscount Milton. Meanwhile, the London Geological Society also attracted prominent ornaments of the great Whig aristocracy, including the Marquess of Lansdowne and the Duke of Devonshire.

These were men coming together on the basis that individuals acting in combination might add to the general stock of knowledge, without prejudice to the very different theories of the earth that they might hold, and without any partiality to their own individual claims to status and prominence.[86] Clearly, they had disparate social and economic interests. It was assumed that there would be a wide degree of latitude in their standpoints, and that therefore authors were to take responsibility for the particular facts and opinions contained in their papers. Few of the statesmen actually contributed articles, but what was important was their public affiliation with and subscription to this model of cooperation. Variety was tolerated in the name of the overarching principle of advancement through factual accumulation. The ostentatious evocation of this foundational principle, rather than any overt politicisation, was characteristic of the intellectual coalition. In the five volumes of the first series of the *Transactions* there is only one article with a pronounced political-economic bias, a polemic against the support of mining by national expenditure.[87] More typical was Henry Grey Bennet's expressly plain and unpoetic account of the peak of Tenerife (in the context of Milton's "Fiery surge").[88] Factual ostentation was not only a methodological nod in the direction of Bacon, but the groundwork for intellectual sociability.

Lord John Russell skilfully appropriated the narrative of geological partisanship, transforming the debate into a template of intellectual coalition-making:

> On the subject of education there appears to me to have been a change similar to that which took place many years ago on the subject of geology. At that period geologists were divided into Neptunians or Vulcanians [*sic*], Wernerians or Huttonians, & hot was the dispute regarding the best theory of the formation of the crust of the

> world. Some of the wise men said, however, "Let us first investigate the facts, without troubling ourselves what theory they may confirm or validate." This has now been done for many years, & assuredly while controversy has diminished science has gained by the change.[89]

'In like manner', claimed Russell, 'popular or national education has been a matter of warm contention among sects or parties till the present year'.[90] The acquiescence of the government in the appointment of a Royal Commission (the Newcastle Commission) at the prompting of the National Association for the Promotion of Social Science disclosed a comparable provisional trajectory. Neither the Wernerian, nor Huttonian, nor Anglican, nor Nonconformist partisan was expected to commence by amending his fundamental conception. The style of this compact, illustrated by the geological template, was both judicial, or arbitrational, and irenic. Reform borrowed the prestige of 'facts', established a factual tribunal and thereby forged a mode of provisional reconciliation. Russell's own geological views are elusive. His portrayal of party conflict and the way it is transcended by knowledge points to a feature of political style rather than a geological ideology.

If the parallel between the benighted neptunist, or vulcanist, and the embittered Nonconformist, or Anglican, is indicated in this example; the role of reform is assimilated to that of the facts themselves (or rather, to the process of their elucidation). The provisional character of the process underwrote its progressiveness, and was of course itself a deeply political manoeuvre. This claim to transparency, to a politics elided with enquiry, was a recurring move by Whig–Liberals. It was present in Lord Lansdowne's platitude that the basis of reform lay in the exploration of concrete conditions according to philosophic principles,[91] and in Palmerston's assertion that, amid the collision of opinions, 'the object of all science is truth, & the science of Government is an investigation of truth'.[92] Science offered a process of prejudice tempered by enquiry leading to improvement, allowing Reform politics to locate itself on the latter segment of this narrative line. In dramatising the pattern, the course of geological debate could be conscripted in a stylised manner, confirming the presiding presence of Whig-Liberalism both over and outside political machination. The invocation of natural science contributed to a curious inversion of Whiggish politics, which had hitherto been encumbered with the stamp of factious opposition. In this case, scientific analogy facilitated the claim to transcendence of Russell's style of politics.

The intellectual or cognitive coalition was a means by which Whigs could reach out across political fissures, and mobilise a particular style of factual enquiry that bolstered their political identity while being ostentatiously non-political. The intellectual coalition could work with liberal-minded conservatives and Peelites, which explains its staying power in the 1850s, but it could not function in conjunction with the 'passionate' politics of the radicals or, ultimately, with Gladstone. 'Principle' set formations in motion as an enabling condition of sociability rather than in the first place a methodological theory or description of the world. With such manipulation in mind, a parallel can be drawn between intellectual coalitionism and the Whig reception of Baconianism, which was of central importance.

Bacon, science and sociability

Thus far we have treated rational sociability as a style of portrayal assimilated to the assumptions of party and coalition. One contest in particular in the field of ideas was particularly relevant to the conception of a sociable, scientific liberal style: the argument over the legacy of Sir Francis Bacon, Baron Verulam. It is appropriate to turn to the intellectual background of this debate, even though it was at some distance from practical politics. Bacon was a reference point for Lansdowne, Horner and Brougham. In the Whig appropriation of Baconianism, epistemological collaboration and social utility mattered quite as much as inductive method.

The sympathisers of Rockinghamite and then Foxite Whiggery maintained a sustained, if intermittent engagement with the Baconian canon, in order to signify opposition to traits of which the Whigs and later Liberals disapproved, and their contrasting adherence to mitigated reform. A useful benchmark is the invocation of Bacon in 1794 by Edmund Burke (1729–1797) during the impeachment of Warren Hastings. The Lord Chancellor, according to Burke, had been a model of contrition before his parliamentary accusers in his own impeachment. The contrast with the unrepentant Hastings was severe:

> We know from history and the records of this House, that a Lord Bacon has been before you. Who is there that, upon hearing his name, does not instantly recognise everything of genius, the most profound, everything of literature, the most extensive, everything of discovery, the most penetrating, everything of observation on human life the most distinguishing and refined?[93]

That is, a figure, who dramatised both Hastings' lack of the same qualities and the inexorable nature of parliamentary justice. The manoeuvre gained its dramatic effect from a fault-line that always threatened to undermine Rockinghamite and Foxite uses of Bacon. The legal predecessor of the Good Old Cause, the cause of Hampden and Sydney, was Sir Thomas Coke, an opponent of Bacon. Bacon himself had plainly been a court intriguer, a 'lion' of the throne of James I. The servility, corruption and royal authoritarianism associated with Bacon implied a commonwealth or country critique of the statesman. The solution was to categorise these as personal weaknesses, of which Bacon became aware and was repentant. They were separated from his meliorist programme, encyclopaedic knowledge and embrace of free and sociable enquiry. It is the last of these aspects that is of particular interest in the present context.

Susan Cannon argued nearly thirty years ago that Baconianism had little to do with the actual practice of nineteenth-century science. Richard Yeo, by contrast, restated the importance of Baconian images in natural scientific apologetics, at least until the deterioration in Bacon's philosophical reputation precipitated by Brewster and Liebig amongst others.[94] Holding this view, Yeo also believes that the purely political deployment of Bacon's name in nineteenth-century Britain 'was neither extensive nor radical'.[95] His argument is certainly borne out by the brief political examples that he adduces. The moderate Whig historian Sir James Mackintosh urged Basil Montagu to abandon his interest in William Godwin in order to edit a new edition of Bacon's works. The evangelical divine Thomas Chalmers praised the careful induction of social data. And William Gladstone concluded, after a lengthy perusal of the subject in 1834, that the *Novum Organum* permitted spiritual elasticity. Meanwhile, the radical publisher Richard Carlile numbered Bacon, along with Newton and Locke, among the patrons of intellectual superstition.[96]

However, Carlile's assessment points to the broader significance of Bacon as an intellectual weathervane. Pocock and Winch describe a shift that occurred over the course of the nineteenth century in the characterisation of philistinism.[97] Redescribed by Burke, Blake, Coleridge, Carlyle, Matthew Arnold and ultimately William Morris, the polite culture of commercial society became the vulgar materialism of the manufacturing age. This manoeuvre, signifying the overwhelming of civic humanist republicanism by political economy, was accompanied by the myth that presiding rationalism had been dominated by the mechanical philosophy derived from Bacon, Hobbes, Locke and

Newton. In the eyes of Romantic reaction, such apostles of mechanism had anticipated the administrative approaches developed by Condorcet, Hartley and Bentham which amounted to a highly reductionist science of legislation.[98] The distance between Whiggery and Benthamism was elided, so that the attack on the mechanical quadrilateral (Bacon, Hobbes, Locke and Newton) could become an assault on a dominant Whig culture.[99]

In this context, it is significant that the most distinctive Whig defence of Bacon was not of his inductivism, but of the cognitive sociability that he supposedly represented. One aspect of this was the gospel of use, embracing the familiar and often-analysed discourse of epistemological improvement, and its corollary of open demonstration.[100] This kind of Whiggish advancement was encapsulated by the London Mechanics' Institution founded in 1824 by George Birkbeck (1776–1841). On laying the foundation stone of the institution, Birkbeck had declared that the aphorism of the 'illustrious Bacon' had been refined: 'knowledge is wealth – is comfort – is security – is enjoyment – is happiness.' Henry Brougham dedicated his *Practical Observations upon the Education of the People* (1825) to Birkbeck, just as the SDUK began operating under his aegis.[101] A little later, Macaulay dismissed Bacon's inductivism, but praised him for giving a social motive for discovery.[102] While a Stoic argues that death is no evil, the Baconian sets out to vaccinate.[103] The use of Bacon as a utilitarian muse, with the optional baggage of Benthamism, facilitated the elision of mechanical politics and the Whig tradition mentioned above.

Macaulay wrote on Bacon in 1837, after a series of developments in experimental theory had weakened the automatic deference to the inductive ladder.[104] Nevertheless, there was a second aspect of Baconian sociability, coterminous with Bacon's method, which could be maintained as a moral standard even while the methodological framework was amended. This was the collegiate mode of enquiry that Bacon was held to have advocated. Lawrence Goldman has shown how sociable scientific research provided an institutional framework for the reproduction of mid-nineteenth-century liberalism.[105] Even Macvey Napier (1776–1847), who robustly defended the leading principles of the *Novum Organum* and their applicability to mental and moral philosophy, counted as one of Bacon's chief legacies the collaborative work of philosophical societies.[106] This Whig discussion of Bacon privileged him over Newton, who cut a somewhat isolated figure. The most succinct and widely disseminated portrait of Baconian sociability was penned by Professor Dugald Stewart in his 'Preliminary Dissertation' to the

Encyclopaedia Britannica, first issued in the Supplement of 1815–1824, and republished in a revised form under Napier's editorship.[107]

Stewart dramatised the sociable Bacon by contrasting him with the solitary Hobbes. Hobbes had gained his positive attributes – his disregard for scholasticism and bold spirit of enquiry – from Lord Verulam, whose circle he had joined after returning from his travels in France and Italy. Happy, wrote Stewart, were Hobbes

> if he had, at the same time, imbibed some portion of that love of truth and zeal for the advancement of knowledge, which seem to have been Bacon's ruling passions. But such was the obstinacy of his temper, and his overweening self-conceit, that, instead of cooperating with Bacon in the execution of his magnificent design, he resolved to rear, on a foundation exclusively his own, a complete structure both of moral and Physical Science; disdaining to avail himself even of the materials collected by his predecessors, and treating the *experimentarian* philosophers as objects only of contempt and scorn![108]

Hobbesian philosophy instead became the medium of deductive folly. Hobbes transgressed against all the Whig virtues. He cut himself off from a sociable intellectual community, the party of Baconian scientists attached to the Royal Society. Rejecting the fruits of systematic experience, his speculative reason produced an abstract justification for tyranny. His manner was mocking and iconoclastic, and contrary to collaborative construction. Furthermore, Hobbes' anarchic individualism provided a natural counterpart to the extreme authoritarianism of his theoretical scheme. The comparison with Hobbes reinforced the conservative and pragmatic temper with which Bacon, according to Stewart, had recommended the prosecution of scientific politics. In the eighth book of the *Advancement of Science* Stewart wrote,

> The science of such matters certainly belongs more particularly to the province of men who, by habits of public business, have been led to take a comprehensive survey of the social order; of the interests of the community at large; of the manners of natives; and who are thus prepared to reason concerning the wisdom of laws, both from considerations of justice and of policy.[109]

Such scientific statesmen would be in a position to appreciate Bacon's cautious and temperate maxims on the subject of political innovation.

On the one hand, 'a stubborn retention of customs is a turbulent thing, not less than the introduction of new'. On the other, 'Time is the greatest innovator; shall we then not imitate time which innovates so silently as to mock the sense?'[110] Accompanying these aphorisms are the 'profound reflections' in the first book of the Advancement of Science

> on the necessity of accommodating every new institution to the character and circumstances of the people for whom it is intended; and on the peculiar danger which literary men run of overlooking this consideration, from the familiar acquaintance they acquire, in the course of their early studies, with the ideas and sentiments of the classics.[111]

Here were two versions of damaging abstraction, or literary politics, which were absolutely central to Whig critiques of theory. The first version was individual speculative innovation, and the second was the unmodified reproduction of classical models. The French Revolution epitomised both types of abstract error. In Bacon's superiority to Hobbes and in his objection to classical republicanism, Philosophical Whigs found a rhetorical archetype for this species of double critique. As a model, Bacon understood the importance of gradual political development and of choosing one's moment; his politics were cognitive rather than contractarian, and his favoured statesmen knew much about other societies as well as the history of their own.

This contested terrain over Bacon's legacy of sociability had only a passing relationship with the details of method, though the question of collegiality was later absorbed into Brewster's claim that Bacon left insufficient room for scientific genius. The initial thrust of the Romantic criticism of Bacon was that he represented an externalised ordering of information about the world:

> *Bacon*: I have persuaded men, and shall persuade them for ages, that I possess a wide range of thought unexplored by others, and first thrown open by me, with many fair inclosures of choice and abstruse knowledge. I have incited them and instructed them to examine all subjects of useful and rational inquiry: few that occurred to me have I myself left untouched or untried: one however hath almost escaped me, and surely one worth the trouble.
>
> *Hooker*: Pray, my lord, if I am guilty of no indiscretion, what may it be?
>
> *Bacon*: Francis Bacon.[112]

This tableau, an encounter between 'Lord Bacon and Richard Hooker', depicted by Walter Savage Landor (1775–1864) in his *Imaginary Conversations* illustrates the anti-Whig view of Bacon as the unreflecting subject, the enquirer without self-knowledge.[113] In turn, this was a component of the gradual discrediting of the humanist and civil-republican ideals upon which enlightened Whiggery drew. It was against this background that Samuel Taylor Coleridge attempted his famous hermeneutical re-engineering of Bacon, seen most clearly in *On the Constitution of the Church and State* (the second edition of 1830). William Blake had depicted Bacon as the first of the Druids.[114] Coleridge, on the other hand, was determined to dissolve the nexus between knowledge and practical utility forged through the definition of knowledge as outer-directed sensualism.[115] Coleridge wished to maintain a very different kind of Baconianism, that of Bacon, the British Plato, whose Ideas in nature were material laws.[116] Laws were the external contemplation of that which was contemplated subjectively as ideas. This was a precondition of the cognitive tasks Coleridge wished to set the National Clerisy. The propagation of a unified civilisation rested on an empirical idealism reminiscent of Vico's: the interpretation of languages and traditions on the basis of an Idealist *prima scientia*.[117] Coleridge's reinterpretation of Bacon can be seen as an attempt to compete for the social aspect of his thought. His programme was to be the collective project of the Clerisy, synthesising exterior and interior knowledge through the new Platonic idealism.

The most sophisticated treatment of the sociability of Bacon's canon was the discussion of the way in which Baconian enquiry had the capacity to stabilise the shared language of a given community. This was a problem that engrossed Francis Horner. One of the models closely associated by Horner with Baconian logic was that found in the *Discourses* of Sir Joshua Reynolds.[118] Greatness could be attained through steps of human effort, recognising the universality of rules of excellence across all arts. Horner termed this 'the economy of intellectual labour'. This concept of Baconian reasoning as a model of mind implied a set of procedures that should be imitated. In 1803, Horner sketched out a personal taxonomy of models for disseminating truth. Bacon was classified with Beccaria, Dugald Stewart, Paley, Adam Smith and David Hume as an authority offering moral reasoning relevant to politics. Copernicus and Galileo, on the other hand, belong to another division, of indirect models of reasoning (models based on empirical physical observation).[119] For Horner, Bacon was classified according to his articulation in literate prose. Baconianism as a model, a philosophy of mind, provided a means of getting to grips with the circulation of terms within the economy of the intellect. Abstract terms, once linked with sensible objects, were then separated through cultural permutations and were exchanged in the world of use by day and by hour. A nominalist analysis prompted by Bacon, according to Horner, would reveal the sensible foundation of general meanings that had gained their own currency. A proper mimesis, in other words, would suggest to the imagination the sensible basis of abstract terms. Horner immediately preceded this discussion with a reflection on the progressive nature of the commercial spirit of money-making, which both mirrored and drove forward the concrete mechanical arts. Horner's outrage at the continuing imposition of a non-convertible currency after 1797 becomes perhaps a little clearer: it was a spurious mimesis of value.[120] Conversely, Baconianism suggested the true circulation of descriptions.[121]

Thus the Baconianism of the early nineteenth century was preoccupied by more than the question of inductive procedure, or even the secular increase of human power.[122] Dugald Stewart's contrast between the cooperative experimentarian and the isolated Hobbes gets us closer to the real sociable resonance of Baconian reference. Thus far this chapter has sought to get past discussions of Whiggish science according to its place on a progressive–conservative scale, its optimism or its teleology. Instead it has sought to show how the Rockinghamite tradition of party opened the way to intellectual coalition-making, and to demonstrate that Scottish Whiggery disclosed a distinctive cluster of sociable and Baconian styles. Next we will move on to look at the liberality debates that framed another dimension of political display.

4
Liberality

Science, Whig politics and the struggle for liberality

Between c. 1790 and 1840 ideological possession of liberality (as defined below) was vigorously contested in British politics. Liberality was especially important for the Whig party, the party of the Foxites and their successors, although they did not manage to dominate it as their political property until the 1830s. At the same time liberality was invoked in the rhetoric and patronage of natural science. Prominent Whigs promoted liberal science, although such scientific liberality was never an exclusive Whig domain. What then was the nature of this claim to political territory?

When John Henry Newman sought to attack the character of mind from which Liberal theology grew up at Oxford around 1820, he offered 'enlightened views, largeness of mind, liberality of sentiment'.[1] The last of these phrases connotes, but goes beyond, the liberty that is generally taken to be the subject of liberalism (or the adjective 'liberal'). Historians discuss political, social and economic liberalism, but there was something else underpinning liberal values, something perhaps not so obvious in policy formation. J.E. Cookson observes that the Friends of Peace (1793–1815) applied liberal enquiry and liberal views to a swathe of topics including religion, education and political economy, opposing liberal principles in all these fields to narrowness, bigotry and prejudice.[2] Liberal enquiry aimed at a comprehensive approach to the truth. Liberal politics emerged within this frame as a devotion to the general good against narrow interests. Cookson argues that it was in the context of the general good that liberality connected with the liberty that individuals were entitled to claim in society.[3] The great preoccupation with civil and religious liberty correlated with, and made sense within, a broader liberality of sentiment that implied generosity and comprehensiveness. This kind of liberality discourse was widespread among the Foxite Whigs. Sir James Mackintosh told Lord Holland in 1805 that whatever was liberal was generous.[4] In 1817 the Duke of Bedford linked conciliation and kindness in government with preservation of rights to petitioning and reform of abuses.[5] The liberal paternalist, activist aristocratic ethos of the 1830s described by Peter Mandler represented the mature development of such Whig political liberality.[6]

This Foxite version of liberality, favouring political reform and opposing war, was challenged from different directions. Abraham Kriegel notes Burke's argument in his *Reflections* (1790) that without generosity of spirit, liberty was not liberal.[7] The illiberal liberty condemned by Burke belonged to the French National Assembly, and it was not in the first place their lack of generosity, but rather their coarseness and vulgarity that prompted the gibe.[8] The spirit of religion and gentility made the difference between liberal minds and a gross, stupid nation. The liberty of the assembly was not liberal and its science was presumption.[9] J.C.D. Clark reminds us that the contemporary meaning of 'liberal' (following Samuel Johnson) was 'not mean; not low in birth; not low in mind...becoming a gentleman...munificent; generous; bountiful; not parcimonious' [sic].[10] In the 1790s genteel liberality could readily be appropriated by the opponents of reform. Under this Burkean construction the liberal spirit was attached to a society of piety and rank, implying something closer to chivalry. Donald Winch points out that it was the chivalric argument of the *Reflections* that attracted Edward Gibbon, being redolent of the Scottish philosophic histories that saw politeness and gentility emerging out of feudal society.[11] Burke's melodramatic portrayal of the sufferings of the Queen of France marked the rhetorical struggle for liberality, implying the illiberal baseness of the revolution and its sympathisers. Of course for those among the Rockinghamites who followed Fox in the early 1790s, generosity in bestowing political rights and in granting civil and religious freedom were both examples of aristocratic chivalry and condescension, but for those who followed Burke the fear of Jacobin meanness could give rise to a rival and politically reactionary form of liberality.

Thus Clark is mistaken when he argues that 'liberal' had no specifically political connotation until the borrowing of the label from the Spanish *Liberales* in the 1810s.[12] There was a struggle to mobilise liberality in different directions rather earlier. As Clark and Cookson both explain, the virtues of Christian charity and generosity were invariably claimed by religious Dissenters,[13] but churchmen had little difficulty in harnessing the virtue of liberality to their own causes, notably law, order and social calm.[14] In February 1798 the Reverend Samuel Glasse (1734–1812) preached on *National Liberality and National Reform Recommended*.[15] Glasse was a solid Pittite and loyalist churchman, a chaplain in ordinary to George III, a member of the Royal Society and advocate for numerous philanthropic endeavours.[16] His sermon was an exhortation to support the wartime government at a time of financial difficulties. A year previously the Bank of England had suspended cash payments, forcing Pitt to treble the assessed taxes on luxuries and then in December 1798 to introduce an income tax. Taking as his theme 'Render therefore unto Caesar', Glasse urged his flock to avoid the sin of the Pharisees, who were averse to paying tribute, and the sin of the Herodians, who forgot their religion in ingratiating themselves with the civil power.[17] Public service and sacrifice were necessary for the defence of the constitution: 'The present situation of our affairs calls loudly upon all ranks and degrees of persons among us, not merely to contribute to the stock of National Supplies, but, in an eminent degree, to contribute towards the stock of National Piety.'[18] Liberality meant generosity towards the needs of the state and the giving of pious service. More particularly it meant national reform in the moral transformative sense of the evangelical revival: 'Let us not wait until judgments *drive* us to a general *reformation*; but let the goodness of God, of which our Nation hath so abundantly partaken, *lead* us to *repentance*.'[19] Under this loyal liberality, national gifts were spiritual as well as material, which implied that the state was more properly the recipient than the source of political and economic concession.

It is true that in the post-Napoleonic period, Anglican hostility to 'plans of reform' (that is, to structural, political or constitutional change) was sometimes expressed as outright condemnation of liberality – a sign perhaps of the success of opposition Whigs in claiming that concept. But this was invariably defined by conservatives as 'modern' or 'spurious' liberality, with the implication that a 'true' Christian liberality could be retrieved. In 1819 the Reverend George Preston gave a visitation sermon, *The Principles of Modern Liberality and Fanaticism Inconsistent with the Simplicity of Gospel Truth*.[20] Preston argued that the complexity of the 'crude fantastic theories of modern liberality', where 'every pretender has his own plan of reform', was incompatible with Christian simplicity.[21] Liberality – in modern phraseology – meant licence, lukewarmness, indifference to faith and the renunciation of all authority and discipline.[22] Towards the end of the 1820s, amid controversy over the Test and Corporation Acts and Catholic disabilities, one anti-Catholic Emancipation tract was caught between the need to vindicate 'English liberality' and the desire to decry 'the very word, liberality, [which] according to the present acceptation of it, would destroy the foundation of society'.[23] In 1830, William Sharpe, the chaplain to the High Sheriff of Somerset, complained about the difficulty of asserting post-pentecostal discrimination and difference for fear of the charge of bigotry. Pentecost had divided the nations, but one had difficulty in preaching this fact. This 'unwarrantable and extravagant extension of the doctrine of Christian charity' held its power in the tone of conversation, in literature and in 'popular deliberations'.[24] Put directly, some clerics felt hampered by a public culture of intellectual generosity. They felt trumped by displays of such generosity, and felt that they were losing rhetorical ownership of a charitable virtue they wanted to claim and define in their own terms. It was not exactly an ideological problem, in the sense that 'false liberality' posed unanswerable theological questions; rather it was a problem of public mood and demeanour.

Sharpe's lament shows how the spirit-of-the-age discourse of reformers had by the turn of the 1830s penetrated the assumptions of their opponents. Liberality now denoted for both sides an atmosphere of political reform and lessening of religious division. Thus it seemed to an Anglican apologist in 1828 that a

> *spirit of spurious liberality is predominant*... Nor was it by such lax conduct that the State and Church of Britain have become the admiration and envy of the world. But liberality will be her destruction, *I mean modern liberality, not Christian Charity, whose appearance modern liberality attempts to assume*, but it will not do.[25]

By the following decade the language of liberality was beginning to feel old-fashioned, and Anglican polemical tracts in this genre began to wane. For example, in 1841 the Bishop of Llandaff Edward Copleston (1776–1849) presented a gentler version of the false liberality argument. According to Copleston, liberality was only one ingredient of Christian charity, and genuine liberality lay in tolerating differences whose distinct value was admitted. False liberality was indifferent to the truth of competing doctrines, and was therefore worthless. The variation in tone might be rather due to the fact that Copleston's earlier political orientation lay towards the Grenvillites and liberal tories, rather than high ultra-toryism. Just as it was possible for leading liberal tories to join the Whigs in 1830, so perhaps it was possible for liberal tory intellectuals to accommodate to reforming liberality. Coplestone's argument indicates that the 'false liberality' thesis had retreated to the purely theological domain after the key battles over Emancipation had been lost.

We are looking, then, at a process of rhetorical encroachment through which the Whig reformers gained dominance over the language of liberality. Yet they never acquired a monopoly, even in the 1830s when Whig ownership of liberal politics had become more plausible. It is now necessary to take a closer look at the structure and sources of Whiggish discourse concerning liberality. The Machiavellian or neo-roman canon may seem to offer a plausible provenance, given that a free state based on virtue and independence might be supposed to draw on civil sacrifice and generosity. Whig statesmen were demonstrably reading the Florentine author; for example, in 1822 Lord Lansdowne told Stanley to read Machiavelli's discourses on Livy,[26] while later civic republicans such as Algernon Sidney and Lord William Russell were also among the venerated luminaries of the Whig pantheon.[27] However, there was no simple one-to-one relationship between Machiavellian argument and liberality. Just how the two discourses related to each other can be illustrated in a very specific example advanced by Richard Brent as evidence of Whig Machiavellianism: Lord John Russell's historical writing and advocacy of parliamentary reform.

Brent points out that in the peroration of the speech with which he introduced the first Reform Bill (March 1831), Russell included a portrayal of the virtuous aristocrat, one who relieved the poor and performed important local and social duties.[28] And undoubtedly, invocations of public generosity, munificence and charity were essential aspects of Whig rhetoric at the height of political battle.

> The gentlemen of England have never been found wanting in any great crisis. When the country was engaged in war against the national enemy – when the honour and security of the country were assailed – they were ever foremost. When burthens were to be borne, they were ever as ready to bear their share as any other class of the community. I ask them now, when a great sacrifice is to be made, to show their generosity – to convince the people of their public spirit – and to identify themselves for the future with the people.[29]

Generous sacrifice was a sign of public virtue, and martial selflessness a reformist version of Burkean chivalry. Brent suggests that Russell's Machiavellian constitutional moralism had been adumbrated in various literary works and in an earlier speech on parliamentary reform in April 1822.[30] Russell had stated then that 'Machiavel may be styled the Whig commentator on Roman history.'[31] Yet when we attend to the precise context of that remark, we find that it was made as part of a debating point against Henry Bankes (1757–1834), a rotten borough member and patron of Corfe Castle (1780–1826) and an opponent of reform. Bankes, who had had published *The Civil and Constitutional History of Rome from its Foundation to Augustus* in 1818, was the 'Tory commentator' on Roman history just as Machiavelli was the Whig one. It was evidently Bankes who had made the Machiavellian argument, and Russell who was turning the tables. In fact Russell's extended classical argument at this point was drawn directly from Livy,[32] and centred on how the aristocrat Appius Claudius restricted the consulship to the patricians, and how the Senate and dictator Camillus reversed the policy with no ill effects. A full version of this story is not to be found in the *Discourses*, and it was in any case a highly tendentious exposition, since Camillus and the Senate really accepted the reverse as a defeat. However, the important point for present purposes is that it was Russell's tendentious reading of Livy, not of Machiavelli.

To spell all this out, Russell was drawing on Livy here rather than Machiavelli, which Brent has previously identified as the source for this section in the speech. It looks as if Machiavelli was the main source because Russell himself refers to the Florentine. However, Russell's reference to Machiavelli was really in riposte to Bankes, and in context it was a rhetorical turning of Bankes' Machiavellian argument. The example shows that one has to be careful in diagnosing civic republicanism where apparent references to Machiavelli are made.

Brent's general argument about Machiavelli's influence on Russell is convincing with regard to the connection between freedom and civic virtue. Traces can be seen in both *An Essay on the History of the English Government and Constitution* of 1821 and *The Causes of the French Revolution* of 1832.[33] However, on the particular point of liberality the issue is more uncertain. Some aspects of Russell's reforming discourse fit the Machiavellian mould quite well: the comparison between the aristocratic gift of martial service and the gift of political rights, for example, and Russell's insistence on the prudence of concession ('a question whether or not the Constitution would now perish if Reform be deferred').[34] After all, Machiavelli had recommended that 'a Republic or a Prince should ostensibly do out of Generosity what Necessity constrains them to do'.[35] This was one of the polemical points made by Russell against Bankes in 1822, that is that he should have realised that it was prudent to grant what could no longer be refused.[36] However, despite Russell's deployment of Machiavellian dicta for rhetorical purposes, the fact is that the type of Whig liberality he espoused depended on so much more than ostensible and prudent concession. It was a substantive good rather than a mere device by which to accomplish other patriotic ends, however worthy the latter might be. In one important respect, in fact, it was diametrically opposed to the Machiavellian canon, which tended to frown upon altruistic sacrifice or display, fearing profligacy and injudicious loss as a result.[37] If Brent's larger thesis is correct, that is that Russell was working his way towards a form of constitutional *moralism* by the end of the 1820s, then Machiavelli had, at the very least, to be supplemented and tempered by additional moral theory or theology. To be fair, Brent himself argues this when he places Russell in the context of an emerging Liberal Anglicanism that espoused both civil liberty and generosity of the state towards different religious beliefs through a comprehensive church policy (for example in education). With regard to Russell's rhetorical tactics during the 1831 Reform Bill debate, perhaps it would be more accurate to say that he felt himself to be so thoroughly immersed in the Machiavellian tradition, in its problems and original sources, that he was able to argue with or against the grain selectively.

In fact, prudential Machiavellianism, properly understood, is hard to square with Liberal Anglican moralism, which is why the theories of Machiavelli and neo-roman prudential liberality are more appropriate to the earlier Foxite phase of Russell's political development, that is before his transition to Liberal Anglicanism as depicted by Brent. It is no surprise, therefore, that the strongest evidence of Machiavellianism comes in Russell's literary work of the early 1820s.[38] As adduced by Brent, John Burrow has commented that it is the political sociology of the *History of the English Government and Constitution* that displays the clearest civic humanism.[39] There, it serves to frame a historical problem in republican terms. How had England managed to combine liberty with opulence and the expansion of the state? The answer lay in the insular, prescriptive inheritance of civil rights through the generations. Yet even here there were limits. Like Charles James Fox, Russell was very far from accepting neo-classical republican patriotic rhetoric in the French Revolutionary mode.[40]

Quite clearly, neo-romanism was falling out of intellectual fashion in Whig as in wider circles by the later 1820s. Scottish philosophical history had already undermined the Polybian–Machiavellian cyclical historical view, and Macaulay was eloquent in his rejection of 'country' patriotism.[41] Meanwhile, renewed interest in Greek history to some extent displaced the early modern focus on Rome as the central case study of growth and corruption.[42] Macaulay, who wrote extensively on Machiavelli in 1827, treated the Florentine with cautious detachment. Admittedly, Maurizio Viroli presents Macaulay's reassessment as rehabilitation,[43] and it is true that Macaulay sought to rescue Machiavelli from popular demonisation, yet his way of doing this was to place Machiavelli and the essential and undeniable fact of his ruthlessness firmly in his own time and context:

> On the peculiar immorality which has rendered the *Prince* unpopular, and which is almost equally discernible in the *Discourses*... We have attempted to show that it belonged rather to the age than to the man; that it was a partial taint, and by no means implied general depravity. We cannot, however, deny, that it is a great blemish.[44]

Macaulay was sure that the Machiavellian corpus represented a unified whole; that it was futile to try to distinguish between the virtuous Florentine of the *Discourses* and the schemer of the *Prince*; that the Italian Renaissance statesman typified by Machiavelli was half divinity and half snake (to adopt Miltonic language); and that the ancient liberty of the free state endorsed by Machiavelli was a mischievous error. 'The great principle, that societies and laws exist only for the purposes of increasing the sum of private happiness is not recognised with sufficient clearness.'[45] Admittedly, Macaulay had a more sceptical attitude towards neo-roman patriotism than Russell and other aristocratic Whigs. However, it is fairly evident that on its own, neo-roman republicanism and the prudential liberality associated with it could not adequately sustain Whig constitutional moralism. Yet, if Machiavellian liberality did not fully square with Whig liberality, it is necessary to turn to other intellectual traditions in order to explain what was really going on.

This is where the practice of natural science assumes some importance. Here is the Reverend Sydney Smith's description of an idealised Whig existence recounted by Walter Bagehot:

> Why don't they talk over the virtues and excellencies of Lansdowne?... He is full of knowledge and eager for its acquisition.... He looks for talents and qualities among all ranks of men, and adds them to his stock of society, as a botanist does his plants.[46]

Smith's encomium takes for granted the third Marquess of Lansdowne's well-known cultivation of the natural sciences.[47] This was manifested in his membership and vice-presidency of the London Geological

Society (1813 and 1819–1821) and governorship of the Royal Institution (1811–1836).[48] Smith depicts Lansdowne as a public figure who is 'an honest politician, a wise statesman, and has a philosophic mind'.[49] The institutions of gentlemanly science have received extensive attention from historians of science, but what has not been explained is the signification of such interests within the world of high politics.[50] The answer suggested here is that it formed part of the successful Whig attempt to gain dominance over the political style of liberality. The study now turns to the theological terrain, taking as its lead the cluster of argumentative styles that was fostered in the debate over bodies. This discussion illustrated certain Whig positions and tested the tolerance of scientific liberality.

Liberality and corporeality

How did the nineteenth-century party of the Mind (and usually its March) adjust to the theological and political problems posed by rival theories of corporeal bodies? Underlying assumptions about dualism, coherence and uniformity deserve to be investigated more extensively. Debates over bodies offer a fruitful field over which to assess the effect of the background sensibility of scientific liberality. This was not invariably the same as the effect of social and economic liberalism. Boyd Hilton notes that moderate *laissez-faire* – 'liberal' – Whigs and Tories were instinctive mechanistic natural theologians and philosophical dualists: believing, much as Palmerston did in 1829, that mind was the only motive power in nature and everything that was not mind was inert.[51] Such natural dualism had a number of inflections. Thus an intellectual division existed between such liberal-minded natural dualists at the political centre and an incongruous monistic melange of Tory ultras, radicals, pre-millenarians and paternalist Foxites to the right and the left.[52] More generally, a thriving literature has addressed the political and theological echoes of competing corporeal doctrines in the first half of the nineteenth century. In particular, historians of science during the long aftermath of the French Revolution have explored the radical dangers (deist, atheist and democratic) of the varied kinds of monistic materialism.[53] In doing so, however, such historians have overlooked the intellectual dangers to the liberal centre posed by Roman Catholicism, specifically by notions of perceived irrationality and unreason.

The existence of Foxite monistic materialism has already been mentioned, in the context of the patronage of John Allen from 1802 and the cultivation of an irreligious, aestheticised salon at Holland House. The political significance of this ideology lay in Lord Holland's adoption of such ideas as a matter of knowledgeable style. Different political emphases could be signalled by the display of different natural philosophical or metaphysical doctrines, while common political ground could be indicated by shared objects of criticism. For example, different and overlapping emphases were displayed in the philosophical approach of Holland on one side, and Althorp and Brougham on the other. The observation of a Foxite–moderate split along the lines of monism and dualism approximately holds. This division related to Foxite irreligion (on the monist side) and a greater willingness to preserve piety on the part of Althorp and (in an unconventional way) Brougham. Yet it is important to remind ourselves that this difference in opinion and emphasis occurred within a broader Whig consensus, in which both sceptic and moderate rejected unreason linked with Catholicism and absolutism.

There was substantial agreement over the character of these bodies that Whigs believed to be significant, and to which Whigs applied a display of comprehensive and rational knowledge. That is, there was consensus over the *coherence* of the natural and political bodies that Whigs believed made up the world. For example, in the Burkean tradition the political party was compared to a body animated by its principles. An incoherent party was a contradiction in terms; parties were defined by the principles that its members shared. More generally, subscribing to the coherence of nature and politics was a way of showing one's reasonableness. The limits of liberality were tested when intellectual and political positions were advanced that seemed purposefully mysterious or incoherent. The significance of the Durham Letter controversy (which we will come to later in this chapter) was that it mounted a challenge of this nature. Realising this will help to explain why a believer as committed to church comprehension as Russell was behaving in an uncharacteristically illiberal manner towards the Catholic hierarchy, while dialecticians such as Gladstone were able to oppose the restriction of ecclesiastical titles.

Whig aversion to irrational bodies came in different forms. Fear of the Mob was in great part fear of the active bodies of the poor, and as such it balanced suspicion of, and intrigue over, the body of the King. In 1792, the Earl Fitzwilliam warned that the first principle of civil society was protection of the individual versus the multitude, that is protection against physical violence and appropriation.[54] Yet this Whig anti-corporealism was not entirely symmetrical. Fear of the

'Many-Headed Hydra' comprised enmity towards irrational physicality, or brute force directed by mistaken or delusive purposes (hence the often-observed Whig obsession with social educational improvement). It was amenable to the therapy of proper, enlightened governance: in plain form, the claim of avant-garde Foxites to be 'Friends of the People'. An ordered, structured sociable body had the capacity to tame unruly physical ones. In contrast, fear of the public person of the monarch constituted antagonism towards a mystical conflation, or paradoxical incarnation, compounded by the hint of sacral authority. This was not a body that could be controlled; it had to be broken – that is to say, public authority had to be separated from the royal organism. Whig political history documented this process of separation. As Lord John Russell asserted in his discussion of constitutional first principles, 'the sovereignty of England did not reside in the King solely', so that even the Tudor succession had had to be confirmed by parliamentary act.[55]

Paradoxical bodies

It is true that in terms of their political and economic principles 'liberal' Whigs rarely shared the radical-Foxite flirtation with immanent or old-fashioned iatromechanical schema. Thus Lord Althorp was prepared to admit that God might have attached a thinking power to matter, but he did not think that a machine could acquire vitality, even if science advanced enough in chemistry and mechanics to produce a mechanism that replicated an organism in every material structure.[56] Althorp wished to preserve divine discretion in the distribution of mind, soul and vitality (some contingent variability in the coincidence of material bodies and immaterial phenomena) to show the indispensable role of Revelation in the knowledge of God. Meanwhile, Henry Brougham's Cartesian separation of mind and matter existed in tandem with a sufficient and self-justifying natural theology. Both positions contrasted with the Holland House natural-theological scepticism of John Allen, whose mockery of natural theology was accompanied by emergent theorising about the physical processing of mind and vitality.[57]

Yet dualism versus monism and transcendence versus immanence (or indeed vitalism versus iatromechanism as a form of non-transcendent argument) were not the only significant questions that were taken to characterise bodies. For many Whigs there was another overarching issue that had to be taken into account: was a body paradoxical or coherent? Thus L.S. Jacyna notices that Allen reserved special opprobrium for the medieval doctrine of the King's two bodies 'in such a manner that the politic body includes in it the natural body, and the natural body has in it the public body'.[58] This constituted the never-dying, always-present monarch. Jacyna sees here a thread unifying Allen's explanation of the human mind without regard to an immaterial principle, his general account of life without the addition of a vital principle and his denial that any divinity was mingled with kingship.[59] This argument is both acute and justified. However, Jacyna passes over the actual grounds of Allen's objection, his protest that the doctrine of the King's two bodies involved the enquirer in 'vain attempts to reconcile impossibilities'.[60] Specifically, the 'transcendent attributes' of kingship 'are incompatible with our notions of a finite, corporeal and mortal being'.[61] Allen was far from arguing that the only tenable bodies were exclusively material ones. For example, the parson in his station was a *corporation sole* showing continuity in his living. The crux was that 'the law, that converts him into a corporation, requires from him no impossibilities, and ascribes to him no attributes incompatible with his character as a limited and created being'. The parson was bound to preach and pray, but not in every corner of his parish at the same time.[62] By contrast, the simultaneous incorporation of the natural body and 'politic or mystical body' of the king was unacceptable because of the impossibilities that it entailed. Double-bodied king-worship was unreasoning and therefore unpredictable. Belief in such an incoherent superstition could only lead to political madness and fanaticism. Moderates abhorred the mechanical, soulless bodies proposed by radical materialism, but whether moderate or otherwise, Whigs could agree in finding impossible bodies unacceptable. Their aversion to the mystical king-sovereign was not the only manifestation of this stance. There were also wider religious or ideological ramifications.

Holland's revulsion at the incoherence of the Trinity has already been noted. Meanwhile, the enlightened Whig pursuit of Catholic Emancipation from the 1790s onwards should not distract from the anti-Catholicism that pervaded Whig attitudes towards impossible bodies. As Lord Althorp concurred with Brougham, omnipotence could not achieve impossibilities such as making a part larger than the whole or allowing Christ to swallow himself in the institution of the sacraments.[63] Or as Lord John Russell put it, 'The bread and wine used in the sacrament, it is true, were to the researches of chemical science not different from any other bread and wine placed on a table for food or refreshment, but in the minds of Roman Catholics, the real body and blood of Christ.'[64] Transubstantiation was a clear impossibility, involving 'materialism, founded solely upon a very gross theory'. It was therefore necessary to translate the historical operation of the theory into coherent terms. The sacramental body was really inert and chemically stable, and not therefore a miraculous substance. That Catholics believed it to be so was a phenomenon of the Romanist mind. Russell repeatedly emphasised the anti-miraculous character of communion in the *Book of Common Prayer*: 'it being against the truth of Christ's natural body to be at one time in more places than one'.[65] Of course, Whigs were far from alone in professing rejection of Rome. However, the thrust of their rejection was decidedly different from that of radical democracy, High Church Anglican Toryism (whether Peelite or otherwise) or pre-millenarian evangelicalism. Most Whigs were comprehensive churchmen, in the manner described by Richard Brent, rather than anti-clericalists *tout court*.[66] Very few Whigs subscribed to 'high' Catholic doctrines whether in the Church of Rome or that of England, and none deployed apocalyptic arguments against Rome. Nevertheless, for many Whigs, fear of Catholicism, as much as fear of atheism and democracy, bounded the kinds of ontology that were found to be acceptable. In practice this meant an unwillingness to tolerate mysterious and impossible bodies, in addition to rejecting the self-sufficient bodies suggested by mechanical materialism.

The extent of Whig anti-Catholicism should not be underestimated. Lord Lansdowne, for example, was quite prepared to satirise the Eternal City, and to mock one devotee who was a bigot devoted to convents and slavery.[67] Lansdowne described the doctrines of the French Ultras as deranged (specifically, their preparedness to conspire in the name of sacral kingship against the limited constitutionalism granted by Louis XVIII). Macaulay fretted about the ability of Roman Catholicism to harness ignorance and enthusiasm.[68] Such beliefs correlated with aversion towards the alliance between the scholasticism and the Vatican, and suspicion of the Papacy as the original Court power.[69] Both Hilton and Wasson remark on how Althorp and his fellow Young Whigs (coalescing from c. 1809 onwards) absorbed spiritual revivalism and were lampooned as Presbyterians, Puritans and Anabaptists.[70] According to the argument, this was a culture of religiosity alien to the older Foxites. Nevertheless, there was one negative point on which the new Puritans and the old enlightened Foxites entirely agreed, namely hostility towards mummery, superstition and Romanist ideas of authority and the Mass. It was the liberal-erastian Russell who wrote that the principles of the late Elizabethan and Jacobean Puritans had 'led them also to free principles of government; their reason quickly stripped a king of his divinity, and their hearts raised the subject to a level with the sovereign'.[71]

There were in fact two corporeal legacies of Rome that various Whigs found philosophically odious. John Allen was quite clear that the paradoxes of the royal body derived from a Caesarist conceit that had contaminated the robust Germanic commonwealths.[72] Churchmen were the agents of this deformity, both at the collapse of the Empire and in later controversies.[73] In the meantime, they rejected a specifically Catholic 'appeal to the senses' as expressed in ritual and the sacraments (sensualism, for example, that enabled the church to teach that 'bread and wine were converted to flesh and blood').[74] As the principal inheritor of the Foxite tradition after Holland, Russell as much as anyone hated not only the social tyranny of Catholic power, as explored in his fictive works *The Nun of Arrouca* and *Don Carlos* (1822), but also its dogmatic statement of metaphysical incoherence.[75]

There are traces of this position in the most glaring episode of Whig anti-Catholicism in our period, the Durham Letter and Ecclesiastical Titles Act (1850–1851). The moment is of particular interest in the context of the present argument because it divided those Whigs and Peelites at the political centre shown to have shared broadly dualist metaphysical assumptions. Political historians have had great difficulty in explaining the contradiction between the Durham letter and Russell's toleration, perceived by many contemporaries.[76] The difficulty is that, as Norman Gash observed, it was a reversal of political tradition for Russell as a Whig to seek to protect royal supremacy and Anglican monopoly.[77] The contretemps is consequently discussed in terms of a grab for popularity or a cynical attempt to smear political competitors with Tractarianism. Russell himself made the rhetorical claim that in the wake of Catholic Emancipation 'I believe that our powers of resistance to Rome, at the present moment are augmented, because loyal Roman Catholics... hold office, and can be admitted to seats in the Legislature.'[78] Yet he was at pains to commend Catholic erastian precedents: the struggle for Gallican liberties, for example, or the Ricardian Statute of Praemunire.[79] There is no suggestion that the royal supremacy at stake involved a rival divine vicarage centred on the monarch. For Russell, the contest was between, on the one side, the representative sovereignty formerly instanced by the 'bar of France' and immediately pertinent to the Crown-in-Parliament and, on the other, the encroachment of doctrinaire Ultramontanism. Furthermore (and this is where the ontological consideration becomes relevant), the argument was predicated on a claim about the behaviour of 'clerical bodies' in general and the Roman Catholic body in particular, namely their propensity to encroach upon public authority (for example, 'it is of the nature of all ecclesiastical bodies to attempt to trench in on temporal matters').[80] The heart of the dispute was whether it was tenable for the authority of the parliamentary sovereign and the Pontiff to exist simultaneously in the same territorial space.

Russell's case rested on a specifically dualist perception of how political authority actually worked. Political behaviour was impelled by influences and powers acting upon the minds of the people, and the 'influence and power' of the Church of Rome was greater over its communicants than that of other churches. This force was exerted through the ecclesiastical claim to infallibility and 'traditionary influence' manifested in sensual ritual. The Papal challenge to Crown supremacy involved in the granting of territorial titles lay in the dissemination of these 'formidable and awful spiritual powers', which were not available to Wesleyan district superintendents or Scottish Episcopalians. Hence the hysterical tone of the Durham Letter, which alleged the replication of these powers within the very gates of the Church of England, under the guise of Tractarianism ('The honour paid to saints, the claim of infallibility for the Church, the superstitious use of the sign of the cross, the muttering of the liturgy so as to disguise the language in which it is written, the recommendation of auricular confession, and the administration of penance and absolution'). What was at stake was the deformation of the free intellectual life available under the Church of England, guaranteed by erastian sovereignty. It was impossible that both forms of authority should exist under the same claim to jurisdiction over the people of any given province.

This argument was comprehensively refuted by the bulk of the leading Peelites (referred to by contemporaries as such, but more properly post-Peelites after the death of their leader in 1850). Yet it is important to grasp the complexity of the various positions taken. Gladstone, Herbert and the other Peelites agreed that the Roman Catholic Church wielded its power through the minds of its congregants; that is, they tacitly subscribed to the same dualist assumptions as Russell. However, they differed from the Whigs because they did not think that the contradiction between Catholic and public bodies made coexistence, in terms of authority, impossible. Russell was convinced that the paradox of sovereignty could not stand, whereas the Peelites were prepared to tolerate it. A protest to the Vatican might be admissible, but the use of legislation against the unstoppable tide of religious liberty was outrageous. Gladstone's riposte was an early example of what Hilton has called his dialectical moral politics: the dramatic and dynamic invocation of confrontational opposites.[81] In contrast, the Whiggism of the prime minister was not in any sense dialectical, and was particularly sensitive to the paradoxes intimated by Catholic claims.

Whig fear of impossible bodies was not as strong as the fear of materialism. Nevertheless, it was present and had political consequences. It was a fundamentally anti-Catholic strand in Whig thought, and therefore at variance with the politics of toleration and emancipation that shaped conventional Whig approaches to Roman Catholicism and the state. More obviously, it underlay the Whig rejection of king-worship and anti-scholasticism. 'Mystery' was the notion that most effectively united Whigs from right to left in hostility. On most occasions this open, plain and rational style of self-conscious liberality was perfectly compatible with, and contributed towards, inclusive and tolerant church politics. But in the crisis over the Ecclesiastical Titles legislation, cognitive liberality fell out of kilter with institutional liberalism. In this respect, the controversy of 1850–1851 exposed the limitations, and even the obsolescence, of the scientific liberality inherited from the eighteenth century.

Dualism and instinct

Whigs developed a number of contrasting rational theological styles. The contrast between a weak, provisional Unitarianism of the kind preferred by Lord Holland (toleration of a God that might exist, and if so, as a unity separate from self-adjusting nature) and a strong, activist unitarianism (belief in a singular Deity that administered not only through natural law, but through His perpetual presence in the instincts of His creatures and historical providence), constituted a line of demarcation between Holland's Foxite agnostic paternalism, which was indifferent to natural theology, and the dualism of Henry Brougham. Both were heterodox and anti-Trinitarian, but Brougham wanted to construct a new and robust natural religion that would underpin Reform. Brougham's sporadic religiosity of tone and rigid dualism placed him closer to orthodox moderate Anglicans, such as Lord Althorp, although his rejection of the necessity of biblical revelation positioned him between the orthodox moderates and Holland House. The comparison between the positions of Brougham and Althorp will therefore repay further attention.

So far as Althorp and Brougham were concerned, bodily dualism consisted in a number of basic precepts. Commitment to the objectivity and solidity of organised matter, and its discontinuity with animating mind, soul or vitality was the common premise. However, ambiguities arose over the contingency of the division according to God's power, and the placing of individuality and therefore of agency within dualist schema. The central dualist argument was that matter could not be its own agent, yet non-material phenomena did not necessarily conform to each other or to discrete physical organisms. Instinct, the passions and vitality inhabited a nether region between compound bodies and the reasoning and moral mind. Moreover, dualist orders were held to be coherent, uniform and predictable. They preferably allowed for a cautious and tempered improvement. Each of these desiderata had somewhat contrasting implications, and their simultaneous argument led to difficulties. Thus despite an authoritative rhetoric, maintaining whiggish dualism proved to be a precarious business. Continuing effort was required to screen out unwanted conclusions and to defend positions that were intellectually vulnerable.

The tensions adumbrated above are illustrated in a lengthy epistolary debate between Brougham and Althorp over instinct. John Yule discusses the structure of Brougham's *Discourse of Natural Theology* (1835) and its reception in the denominational press,[82] while Robert Richards has focused on published works and Darwin's notebooks in order to make the case that Brougham's theories contributed to Darwin's extended delay in publishing the *Origin of Species*.[83] However, Brougham's argumentation is better understood independently of Darwin's retrospective shadow. The debate is instructive because it is possible to supplement the public argument, stylised by Lord Brougham in his *Dialogues on Instinct*, with the private correspondence carried on between the two Whig peers.

The exchange between Althorp and Brougham on instinct in 1836 was conducted in the wake of a correspondence some twelve months earlier concerning Brougham's *Discourse of Natural Theology* and on the religion of Cicero, specifically that of *De natura deorum*, 'The Nature of the Gods'.[84] Althorp argued that Cicero could not have been as confident of God's existence as he and Brougham, since Cicero had not been acquainted with scriptural revelation. In response, Brougham defended Cicero's belief on the basis of his partiality to the Stoic case in the second book of *De natura* and its conclusion. Brougham commended the first Stoic thesis as a design argument from the divine ordering of the heavenly bodies. Althorp objected that the subsequent arguments from epiphanies, divination and prophecy were very weak. Brougham countered by saying that these superstitions showed that the ancients were familiar with the idea of divine interference in the world, a necessary precondition for natural religion.

The issue was the indispensability of scripture, and its inverse correlate, the sufficiency of natural religion. To keep revelation in the frame, the evangelical Althorp repeatedly argued against the possibility of drawing indisputable conclusions from natural phenomena, and stressed the contingency of natural laws deriving from the free will and agency of God. The world worked by divine and uniform legislation precisely because of the disposition of the legislator. Brougham, on the other hand, was keener to bind the Deity in laws of logical necessity. Omnipotence challenged the imagination, but within limits: a lesser number could not be made equal to a greater or a part larger than the whole.[85] These were necessary rather than contingent boundaries, and the same consideration applied to the mind–matter distinction. Brougham could not see how matter as matter could think or constitute mind.[86] Althorp agreed that God could not perform a self-contradictory act, but denied that the mind–matter distinction was such a case: God might have attached a mental power to matter, and so materialist theories of mind might be true.[87] Althorp did not think it likely, but it was a matter of contingent probability rather than of certain deduction.[88]

Brougham and Althorp differed over whether metaphysical dualism was a contingent or necessary truth. They also cogitated over whether instinct was individuated (like reason or souls, which definitely pertained to persons). To what extent could instinct be invoked as an undifferentiated principle illustrative of divine sanction? The key difference between the two men was over whether instinct existed as a legislated mechanism or a perpetual immanent presence. In the third book and general Scholium of the *Principia*, Brougham had found Newton's proposition that the Deity was 'omnipresent, not *virtually* alone, but substantially... In him all things are contained and moved, but without mutually affecting each other'.[89] Brougham coupled this with the 31st Query to the *Optics*: 'And the instinct of brutes and insects can be the effect of nothing else than the wisdom and skill of a powerful ever-living agent, who being in all places, is more able by his will to move the bodies within his boundless uniform sensorium.'[90] As a result, Brougham concluded, one could regard 'the work of the bee, for instance, as the direct and immediate operation of Divine wisdom and power'.[91]

Here it is useful to view Brougham's argument in the longer perspective of his work. He had earlier contended, in his disquisition on the balance of power, 'that in all animals the passions themselves are implanted for the wisest of purposes; that instinct is the principle to which, more than reason, the preservation of life, and the maintenance of order in the universe must be ascribed; and that national councils may be operating what no foresight could combine, while they appear to be swayed only by prejudice or passion' (1803).[92] Brougham's discussion of the balance of power was self-consciously a 'Newtonising' of Hume's earlier essay on the subject. In fact the Newtonian apparatus is used as a means to reprove Hume for his 'misstatement, in one or two points, of the great system itself, which he appears to treat with disrespect'.[93] This was Hume's contention that the maxim of preserving the balance of power was a matter of common sense and obvious reasoning, so that the men of antiquity could hardly have failed to be aware of it.[94] For Brougham, however, the balance of power was a manifestation of an elaborate politico-natural theology, which was amplified through the course of his later writings.

The operation of instinct resembles a market analogy, by which spontaneous order is generated through a systemic logic. However, Brougham's hidden hand of instinct in national affairs in this early piece was a Newtonian presence, rather than solely a metaphor of order. The elision between the instincts of animals and the passions of men permitted a parallel to be drawn between the cosmic system and the relationships between the international powers. By this theory, the immanent presence manifests itself as a common principle, 'political expediency', which is analogous to gravitation. This 'regulates the mutual actions of the European nations, subjects each to the influence of others, however remote... and confining within narrow limits, whatever deviations may occur in any direction, maintains the order and stability of the vast complicated system'.[95] Brougham claimed for the theory a predictive power. Powers that newly arise fall into the pattern of the state system, and any apparent inconsistencies are the equivalent of the irregularities still yielded by contemporary lack of success 'in discovering the whole extent of the planetary law'.[96] There is an intriguing ambiguity over whether the movement of national instinctual bodies is a specimen of direct divine coordination, or whether national passions produce a self-generating logic of international jealousy and watchfulness. The latter is a more conventional natural theological argument applied to politics, and Brougham's reference to the mutual influence of nations may seem decisive in its favour. However, the argument states that this subjection is due to the preceding disposition of national counsels, implanted by divine guidance. This tilts the early piece closer to the position taken by Brougham in his *Dialogues on Instinct*, that instinct was the product of the direct interposition of the divine hand.

This notion incurred resistance from Althorp. In the *Dialogues on Instinct* he is portrayed by Brougham as warning him (Brougham) against being seduced by the temptation that they both shared: 'the disposition we must have, if possible, to believe in a doctrine which, by exhibiting the finger of God as perpetually moving and working before our eyes, seems to bring us constantly into His presence, as if we saw a perpetual miracle wrought'.[97] While the exalted language is Brougham's, the contention is a reflection of the drier point that Althorp actually made in his correspondence: that it would have been consistent with the general structure of nature for God to have created different species with motives common to each species.[98] What Brougham did in the *Dialogues* was to disconnect this point from Althorp's surrounding argument and to turn it from the component of an alternative thesis into a general expression of theological desire and anxiety.

Althorp argued in his correspondence with Brougham, in February and May 1836, that the attempt to expound a theory of instinct was beset by foundational difficulties. Essentially, the answer to the problem would be predetermined by the definition of instinct adopted and its demarcation from reason in the first instance.[99] In the *Dialogues*, Brougham attempted to get round this objection by initially refusing to define the phenomenon, and subjecting it instead to induction.[100] To reach the 'delightful' doctrine of immediate instinctual interposition required, naturally, the 'strictest steps' of inductive reasoning.[101] These steps proceeded in a distinctive way. Compared to Brougham's earlier articles for the *Edinburgh Review*, the ascension seemed to owe rather more to Butler than to Reid. Brougham claimed that a systematic study of animal instinct revealed want of both power and conscious design in the organism performing instinctive acts. From this set of negative observations, Brougham inferred that both power and design had to be supplied, which supposed the interposition of a powerful and designing being, which could only be God. The grounds for moving from power and design to a powerful designer was a positive analogy between the agency of Man and that of the Deity. In turn, this reprised the argument elaborated in Brougham's 1835 *Discourse of Natural Theology*. The structure of human reason could be observed by introspection and conclusions drawn from this about the rational character of the vastly greater divine intelligence. It was the involvement of mind in instinct that made the legislative explanation unsatisfactory. For how could animals behave instinctively as if they were rational, whereas in enacting the behaviour in question they did not seem to know or vary what they were doing? How could behaviour appear to be rational and simultaneously unselfconscious? Only the direct guidance of a superintending mind could remedy the deficiency.

Althorp offered an alternative view. Instinct was a species of motive unaffected by consideration of future consequences.[102] It usually appeared as part of a compound inducement, or mixed motive, for both humans and animals, along a continuous scale. Taking the example of hunger, Althorp considered that it was composed of both a conscious appreciation of the need to support life and health, and the properly instinctual physical impulse. In general, instinct formed only a part of inducement.[103] What was true for humans was also true, as a matter of degree, for animals. While Brougham began with bees, Althorp started with dogs: the dog follows the scent of the game because it is agreeable to it without any conscious expectation of food. However, it then learns by experience, and it was this mixture of instinct with memory that made most such behaviour a variety of adulterated reason. As for the bees, they neither understood the consequences of the geometry of their cells nor showed evidence of learning cumulatively by experience.[104] Their behaviour might fairly be described as purely instinctual, but only as the extreme limiting case of a scale that stretched from man to insect. Althorp saw this scale as mirroring the analogy of creation. Physical progression from vegetable to man had its obvious parallel in the scale of mental powers.[105]

Clearly, both positions were dualist, in the sense that both implicitly rejected the exhaustive mechanical generation of instinctual and sentient behaviour by the material bodies of animals. Yet they differed sharply on the placing of divine agency. The important point to notice is that Althorp's legislative model actually led to a gradation of behaviours, whereas it was Brougham's insistence on the absolute separation of reason and individual instinct that opened the way for the direct operation of the divine presence in instinctual phenomena (as Althorp put it, 'the direct operation of God applied to individual cases').[106] The correspondence suggests that Althorp was perfectly aware that attributing degrees of consciousness to non-human organisms held theological dangers. The gradation of mental power might imply the possession by animals of a soul, which materialists would probably take as a great concession to their doctrines. However, the mental scale was itself the divinely legislated order, and an expression of God's power. The distinctive immortality of the human soul rested on the free exercise of this power, its certainty deriving from revelation. In the meantime, if it was God's pleasure to create a dog with an immaterial soul that lasted as long as its body existed, then it was certainly in His power to create a man with an immaterial soul that could exist for eternity, even after the death of the human body.[107]

The contrast between Brougham and Althorp was theoretical. It was also a difference of religious tone and sensibility. Brougham's doctrine of direct divine interposition opened the way to an ecstatic natural religion, 'which almost enables us to commune with the Deity, as the Patriarchs did of old'.[108] Wishing to preserve the indispensability of scriptural revelation within a mixed economy of belief, Althorp insisted on a more moderate, standard natural theology. This displayed, again according to Brougham's characterisation, a 'hankering after general laws'. Althorp's natural theological argument followed Paley, both in its preference for beneficent design working through universal legislation and in its moderate, sober tone. However, it was supplemented by a scriptural, evangelical imperative. Brougham by no means rejected all legislative accounts. What the addition of immediate divine agency meant was ardour of belief, emotionalising natural religion. If direct interposition were the true theory, then 'we have an additional reason for devoutly admiring the spectacle which this department of the creation hourly offers to the contemplative mind'.[109] Thus there was an impulse running through Brougham's theology that was activist and providential, rather than simply remote and mechanical. Brougham certainly flirted with the idea that the Deity was more than a prime mover and divine legislator. Radical Newtonianism made Brougham in this respect more a child of the heterodox seventeenth century than of the Augustan Age. On the other hand, it was the evangelical Althorp who sympathised with Brougham's loss of office in September 1835 by assuring him that God's chastisements were painful but merciful and intended for our good.[110] Brougham never replied in equivalent terms. By bringing God into the world in a radical and continuous manner, Brougham attempted to assert a different type of intense devotion, and make it uniform and rational. There was little place for sin or atonement in the Broughamite scheme, which treated the gospels tepidly while being ready to invoke those parts of scripture most conformable with natural religion, such as Psalm 139: 'How excellent is thy name in all the earth . . . '[111] Nevertheless, the ecstatic possibilities of the radically immanent Newtonian deity (the communing of the patriarchs) qualifies Yule's depiction of Brougham's formal diplomacy in his relations with the divine.[112] Brougham and Althorp were therefore both in their various ways dualists, and they both partook of distinctive, intense forms of piety.

These pious manifestations differed from one another, but in their rational and scientifico-theological conversation, Brougham and Althorp shared similar terrain. In their correspondence, and then in Brougham's stylised depiction in the *Dialogues on Instinct*, the shared priorities were the demonstration of 'sound philosophy', the disavowal of self-interest and the rejection of esotericism. Brougham's Newtonian immanent presence was a peculiarly rational form of 'mysticism' (if this word is indeed appropriate) that was perceptible in and deducible from seemingly unconscious purposive behaviour in creation. Althorp was careful to preserve the case that divine fiat had the power to suspend natural legislation and a stable metaphysical order, but assumed that radical discontinuities and absurdities were unlikely. As sketched out in his correspondence, this orthodoxy actually made him more open than Brougham to the potential validity of radical and emergent doctrines (for example, the logical possibility of materialism) including the notion of a gradated mixed constitution of mental powers in nature. The line was a subtle critique of Brougham's attempt to dispense with scriptural revelation and had been winnowed out by the time that Brougham came to write up the exchange in the *Dialogues*. In the literary piece, Althorp is given a stolid, cautious and admonitory voice. Still, the attempted register of the conversation is one of scientific liberality, displaying breadth of information and doctrine in amicable free enquiry.

5
The Georgic Tradition

Pastoralism and the Georgic tradition

The central importance of the land to Whig identity has been underplayed by intellectual historians, whose attention has been caught by metropolitan salons such as Holland House, and the urban, academic and professional contexts of London and the Scottish Enlightenment. The civic loci of the university towns and British capitals also remain the essential arenas of the history of science, especially in the treatment of early-nineteenth-century debates. However, it should be noted that the association between Whiggery and urbanism was fostered by rival forms of landed politics, that is to say by the bemoaners of liberal steam intellect, whether Romantic Tory or Cobbetite radical.[1] The caricature of deracinated Scottish philosophers exulting in manufacturing establishments, the dismal science and the prosperity of Manchester and Glasgow was a staple of anti-Whig rhetoric.[2] Recent critics have had to spend a good deal of effort in countering this tradition and in disentangling Whig thinkers from those whom S.T. Coleridge named as the proselytisers of facts without Ideas, liberals and utilitarians.[3] One way of doing this has been to relocate a Whig tradition that extended through the 'long' eighteenth century and that was distinct from classical liberalism. It rested upon a tension analysed by Pocock, namely that a substantial landed class was forced to manage a growing system of public finance.[4] This Whig tradition brought with it ambivalences relating to the growth of commerce, and then industry. The reappraisal has strengths, but it rests on a fairly narrow (if sophisticated) exploration of civic republicanism, emphasising the superiority of real over moveable property.[5] Other implications of the agrarian basis remain to be ventilated.

The status of statesmanship was transformed not only by an emphasis on the knowledge of the statesman towards the end of the eighteenth century but also by an enhanced pastoral role. The statesman was the leader and guide of the people; he was also the cultivator of laws. The respectability of the statesman was confirmed by his productiveness, his utility, and this was expressed both mechanically and in his management of cultivation. That is, even as Britain witnessed the beginnings of industrial society, a key conceptualisation of the statesman's contribution was agrarian and pastoral in character. It is therefore essential to revaluate what the land and rurality meant to Whig politics, drawing attention to the contribution of a recognisable georgic tradition, before relating this rural orientation to the appropriation of science.

The scaffolding of turn-of-the-century Whiggery was the dynastic structure of great agrarian clans holding estates across Great Britain and Ireland.[6] Despite periodic crises, absolute levels of landowning income continued to grow during the nineteenth century, at least until the 1870s, even as the aristocracy diversified their interests.[7] Returns on the Lansdowne estate in Kerry indicate that annual gross rentals increased by a quarter between 1855 and 1864 (the final decade of the third Marquess), while net remittances reached approximately £5000 per annum.[8] Agriculture was the predominant but not invariably the sole source of dynastic wealth. As E.A. Wasson has shown, the Spencer fortune was dominated by the rentals from 20,000 Northamptonshire acres.[9] However, up to a fifth of Fitzwilliam's income in the 1820s came from collieries and other industrial interests in the West Riding.[10] Clearly, then, landholding was not the only economic basis for the political formation studied here, nor was agriculture always dominant within the landed economy.

Of course, not every improving landlord was whiggish in politics, nor did every Whig notable hold land, though the political identity subscribed to by even these Whigs was heavily coloured by enlightened estate management. On closer examination we find a range of articulation, that is to say a number of different ways of projecting the land as a governing nexus of values. It is by one particular set of values that we should understand the agrarian basis of Whig statesmanship rather than by a doctrine of economic determinism. A vision of the landscape grounded these values.

One idealisation of rurality that has already been noted by political historians was the comparison between estate management and the operation of public authority. Here, society was an extension of the private estate, resting on cooperative, harmonious and unequal power relationships.[11] More acutely, the pastoral aspect of Whig statesmanship was indicated by its stylised idyllic environment. In his *Six Months Tour through the North of England* (1770), the agriculturalist Arthur Young praised the Rockingham estate at Wentworth House in arcadian terms drawn from Zuccarelli and Poussin.[12] Earl Fitzwilliam (1748–1833) inherited this elegant landscape on the death of his uncle, the Marquess of Rockingham, in 1782. The affectation of the enlightened rural idyll was noticed by contemporary critics. One caustic item in the *Times* of 27 August 1790 observed that Burke spent his summers practising experimental husbandry and translating Virgil to local farmers who did not understand him at all.[13] This satire was directed against a recognisable literary and stylistic canon, the original source of which was Virgil's bucolic poetry, or *Eclogues* (which, together with his *Georgics*, comprised the Latin continuation, in the first century before Christ, of the Greek idyllic tradition).[14] Virgilian texts were often placed at the head of a 'Geoponic' category of writing, including Pliny's Natural History, and the rustic texts of Cato, Columella, Palladius and Marcus Varro.[15] The seminal date in the popularisation of the Georgics was 1697, the year of Dryden's translation. The subsequent fashion for imitative bucolic poetry in the eighteenth century has been well documented, as have the agricultural controversies over the efficacy of Virgilian methods.[16] It is worth remembering, however, that such writing had considerable longevity in a minor way. Bucolic poetry persisted despite the upheaval in taste promoted by the Romantics. As late as autumn 1838, William Gladstone pondered the merits of cross-ploughing in his diary of a Sicilian Tour. Gladstone censored the local peasants for using a plough that was cruder than Virgil's, lacking both mould boards and share-beam.[17]

Despite their interpenetration, Virgilian geoponics was a separate genealogy to the classical republican canon analysed by Pocock and others, and depicted a generally harmonious and practical view of human labour (or husbandry) in nature. But the classical authorities were not studied in isolation; rather they were part of a live tradition whose filtration through more contemporary texts determined their continuing application. This way of treating georgic themes, which was occasionally set in opposition to the purely aesthetic use of the pastoral, had a double usage: not only in the sciences of legislation and political economy but also in analyses of natural scientific enquiry. Virgil was regarded as a teacher in this wider intellectual sense, as well as of specific farming practices.

However, arcadianism was plainly open to charges of impracticability. It exposed vulnerabilities at the same time as it invoked pastoral identities. The specific Whig use of Virgil also had to navigate certain political problems. The background of the arcadian poetry was the decay of the Roman Republic and the dislocation and portents of civil war. The eulogy (indeed the deification) of Julius Caesar and praise of Octavian, later Augustus, did not sit easily with senatorial models of liberty. As Leslie Mitchell has described, Charles James Fox liked to compare himself to Brutus, and identify Pitt, occasionally, with Augustus.[18] Demosthenes was preferred as a model of rhetoric, and the consul Cicero, especially, as the archetype of statesmanship. Unlike the historians of the Republic, such as Livy, Virgil provided a link between the farmer-smallholder (colonus) and the Augustan empire. Virgil as the apostle of cultivation, and the interpreter of nature, was a far more congenial subject for amendment and revision. By circumventing difficulties specific to Virgil's imperialism, the georgic tradition was rehabilitated as an accepted way of asserting the ideal superiority of the land.

The operation of this kind of ideal superiority is paralleled in pictorial art. Andrew Hemingway shows how landscape painting consolidated its appeal among the middle classes in the post-Napoleonic period, with its ambiguity towards status differentiation and claim to authenticity.[19] The Philosophic Radical *London Magazine* could praise it as the most original production of the British school.[20] The trend showed a re-appropriation, rather than abandonment, of the traditional Virgilian or georgic dyadic opposition between simple rurality on the one hand and metropolitan or courtly sophistication on the other.[21] John Barrell has uncovered the complex patterns of reassurance and repression that configured English pastoral poetry and painting, evoking a happy moral economy while defending the inequality between serene consumers and laborious producers, whose happiness was promised but necessarily deferred. The centrepieces of his analysis are the two pictures that Thomas Gainsborough gave to the fourth Duke of Bedford (1710–1771) in 1755: 'Peasant with Two Horses' and 'Landscape with a Woodcutter Courting a Milkmaid.'[22] Barrell argues that this pattern came under pressure towards the end of the eighteenth century as economic change and fear of shortage subverted the ideal of harmony. The shift can be illustrated by the Malthusian condemnation of Godwin's dreams of ease.[23] The Malthusian concept of the ever-present danger of a subsistence crisis contradicted an idyllic view of the countryside. However, the landscape genre was not discarded – if anything, the simultaneous benevolence and repression of images intensified.[24] It became more important, not less, to show placid country scenes and producers happy with their lot.

The Georgics, Davy and the Bedford connection

Barrell's exploration of mid-eighteenth-century Bedfordian patronage is suggestive because it formed part of the active political and cultural memory of nineteenth-century Whiggery. The heritage was physically incarnated. For John Russell, the Evergreen Drive at Woburn planted by the fourth Duke, and tended by his successors, symbolised political continuity.[25] The harmonious, domesticated image of cultivation provided a matrix for the conscription of nature:

> And without going into an inquiry respecting the commercial advantages to be derived from the knowledge which we are yearly (I may also say daily) acquiring of the growth, and value, and properties of trees, I will content myself with observing that the genus *Pinus* is probably entitled to wonder and admiration beyond all others; and that, at no distant period, we may see the *Cedrus Deodora*, the *Abies Douglasii*, & others of similar grandeur, naturalised and flourishing, among the cedars of Lebanon, in our British forests.[26]

Lord John Russell's citation of his father on the *Coniferae* was not only a memoir of the Evergreen Drive at Woburn, but the preamble to a political narrative. It was 'the honesty, the attachment to his religion, the country habits, the love of home, the activity in rural business and rural sports in which the Duke of Bedford and others of his class delighted [that] preserved the English aristocracy from a flood which swept over half of Europe'.[27] There was no contradiction here between rurality and commerce, since such cultivation was both commercial and conservative. Meanwhile, the pursuit of Natural History united the classes in the 'observation of nature', whether by 'the love of something natural and rural breaking out in the counting house', or through the rural longings of the manufacturing spinners and weavers.[28] Likewise, Russell insisted that Charles Fox himself had loved the warbling of the nightingale, having once bid Grey to 'read through half the Odyssey to find a quotation' attesting to the cheerfulness of its birdsong.[29] Both as a productive site of social order and as an object of study (inevitably quiet, serene and contemplative), pastoral nature figured tenaciously in Whig stylistic politics.

Lord John Russell continued the long-standing connection between Woburn and the world of urbane Foxism.[30] His father John, the sixth Duke of Bedford (1766–1839), was an active Foxite who sat for Tavistock before his elevation to the Lords (1788–1802). When he moved to the

Lords to take up his peerage, Fox moved the writ for the Tavistock seat.[31] John Russell the elder was a member of the Society of Friends of the People, and acted as Lord Lieutenant of Ireland during the Talents Ministry (February 1806–April 1807). On acceding to the dukedom, he maintained the close links with Arthur Young developed by his brother and predecessor Francis, the fifth Duke and famous agriculturalist, whose life had been cut short prematurely.[32] He vowed to follow up the plans for livestock improvement which his sibling had begun.[33] John Russell the elder was energetic in the tight-knit group of agricultural improvers that included Thomas Coke as its principal. Like his brother, the sixth Duke was elected to the honorary membership of the Board of Agriculture, and later financed both fen drainage schemes and the rebuilding of Covent Garden Market.[34] Bedford promoted the sheep shearings and agricultural gatherings at Woburn, Smithfield and Holkham as places of sociable competition and experiment. This involved such activities as managing the delicate business of adjudicating a ploughing match between Coke and Wakefield, and urging Arthur Young to follow up in his *Annals of Agriculture* the sheep experiment which was submitted to breeders at the 1802 Woburn sheep shearing.[35] Bedford's commitment to agricultural improvement also impinged on his Lord Lieutenancy. He encouraged the efforts of the Farming Society of Ireland to gain a charter, and complained that the science of husbandry was badly understood and neglected.[36] Bedford assured the Smithfield Club that he would be happy to resume his presidency on his return from Dublin.[37] Through his sponsorship of the Bedfordian Medal and his prize for the Bath and West society and its papers, Bedford sought to bring before the public a combination of theory and practice, yielding valuable information.[38] He hoped that the science and ingenuity of competitors would soon be exercised on the topic of manures.[39]

Bedford was a fairly typical member of the agricultural Whig gentry and aristocracy. As will become clear, he is worth focusing on because of his agrarian manners and the intersection manifested by his own sponsored experiments and patronage of Humphry Davy, who was a high tory. Davy had gained his place at the Royal Institution under the aegis of Joseph Banks, whom he was to succeed as President of the Royal Society in 1820, and eventually he was to experience similar criticisms of courtly metropolitanism as his sponsor.[40] Nevertheless, it is not quite right to assimilate his politics to Banks'. Davy was prepared to countenance free trade, and to assert his independence despite his move from provincial science to the metropole. On the other hand, it would be wrong to say that Davy's cooperation with the Bedfords represented resistance to Banks' learned empire. The Bedford connection enjoyed good relations with Banks, and the Bedford connection, Davy and Banks presented different variants of a similar agrarian ideology. Agrarian Foxism was flexible and willing to exploit the similarities that existed between Whiggism and some aspects of high tory thought. For example, as we shall see, both Davy and Bedford were strongly anti-Malthusian. Agrarian Whig manners of patronage and publication – more specifically, the form which these manners took – enabled a high degree of ideological exploitation to take place. Bedford was representative of a current of agrarian Whiggery, which has more often been treated from a technical rather than ideological or stylistic point of view.[41] Plainly, it was easiest for Lord John Russell to invoke his own dynastic tradition, but the experiences of the agrarian Whigs were exploitable as the common heritage of the Whig party – a phenomenon most evident in the iconic status afforded to Thomas Coke. Creevey's hailing of 'Our worthy King Tom' is an example of this regard.[42] On a practical note, exploration of the Bedford connection allows us to approach in detail some of the georgic preoccupations of English Whiggery.

To say that the Bedford connection was representative of an important Whig current invites scrutiny of the mode of public representation that the Bedfords actually patronised. Vignettes of the elaborate ceremony and dining that surrounded the Woburn sheep-shearing point to a style of conspicuous improvement.[43] The ducal apex of the affair was emphasised, for example, by the radial arrangement of tables about the person of the duke (and on occasion his guest, the prince regent). There was a definite Foxite accent to the stylisation, beyond broad landowning conviviality. The painting and engraving (1811) by the cattle-modeller George Garrard (1760–1826) of events at Woburn portray a number of Whig worthies in pastoral poses. The Duke's elder son and heir, Francis, the Marquess of Tavistock inspects a Southdown tup. Bedford concentrates on a piece of broadcloth, and other sons (including John Russell) regard a sheep being held and shorn by two husbandmen. The Country Whig MPs John Christian Curwen (1756–1828), Thomas Coke (1754–1842), Charles Western (1767–1844) and William Northey (1753–1826) are variously arrayed, the last looking over Swedish turnips.[44]

N.B. Penny has discussed the material iconography of Foxism, including the ubiquitous Nollekens busts and the symbolic shrine depictions in many of the great Whig houses (Woburn, the Penrhyn estate, Holland House, Holkham). At Woburn, the temple of liberty, affixed to the greenhouse, was commenced by the fifth Duke (who bequeathed £5000 to

Fox) and finished off by the sixth. The polite public were admitted to the temple (by ticket, on Mondays), and were edified from 1815 onwards by Westmacott's external reliefs, which depicted the progress of society from hunter-gathering through pastoralism to enlightened agriculture, trade and the extirpation of slavery.[45] Urbane Foxism, chiefly the Holland House circle, may have had little to do with practical agrarianism, but the idea of an enlightened landed sensibility remained central to its self-definition. Richard Westmacott (1775–1856) enjoyed a close relationship with the sixth Duke of Bedford, and received a number of commissions, including that of the statue to the recently departed fifth Duke, which was erected in Russell Square (1809).[46] The sculpture portrayed the nobleman amid his wheat sheaves, stock and plough. This was Whig agrarianism at the heart of the improving metropolis. It presented an agriculture that formed part of a progressive narrative, not as a bygone sociological stage but as a current and dynamic component of historical reform.

Bedford's enthusiasm for the Evergreen Drive at Woburn was an indication of a deeper interest in practical botany. In 1838 he lobbied his son John Russell to convince Queen Victoria to support the conversion of Kew Gardens into a Royal Botanical Garden comparable to the Paris *Jardin des Plantes*.[47] He sponsored the publication of *Hortus Ericaeus Woburnensis* (1825), *Salictum Woburnense* (1829) and *Pinetum Woburnense* (1839).[48] However, it is the first of this botanical series that is of particular moment to the present argument. *Hortus Gramineus Woburnensis* was originally published in 1816, and a second edition was produced eight years later. The practical work was done by Bedford's head gardener, George Sinclair (d. 1834), but in the light of the Young correspondence it does not seem implausible that, as Sinclair put it, the idea was 'first suggested by His Grace the Duke of Bedford, of ascertaining the nutritive powers of the different grasses, by the aids of chemistry'.[49] The attribution of initiative to the Duke was, of course, *de rigueur*, as was the acknowledgement by Sir Humphry Davy to Bedford for permission to add to the 1813 edition of his *Elements of Agricultural Chemistry* 'the results of the experiments instituted by His Grace upon the quantity of produce afforded by the different grasses'.[50] Nevertheless, the imprimatur of the Duke upon the argument of the *Hortus Gramineus* and Davy's *Elements* was a significant public signal. For both texts (the former borrowing freely from the latter) presented a distinctive set of political and moral arguments.

The experiments conducted on fodder grasses by Sinclair, with the chemical consultancy of Davy (who devised the process to isolate the 'nutritive matter'), have been recycled as an important source for the 'principle of divergence' in the *Origin of Species*.[51] The identification, which caused some broadsheet excitement in January 2002, has concentrated entirely on this post hoc destination.[52] But as the frontispiece of *Hortus Gramineus* indicates, the Bedford experiments – so far from being bolts from the blue or anticipations of future breakthroughs – complemented a stream of agrarian political economy that had a long history. Sinclair's preface is taken directly from Sir Humphry Davy's *Elements*, and it contains a definite replication of the natalist, productive and rational agrarian argument of the latter text.[53] The character of this argument is explored below, but first it is necessary to delineate the particular features of the *Hortus*. Sinclair had inscribed the original 1816 folio to the Duke; the edition of 1824 was dedicated to Thomas Coke, and it is clear that all along the argument had been set out as an address to that worthy. Coke was praised for his 'enlightened and extensive views, vigorous and persevering industry, generous patronage, and liberal policy'.[54] These virtues had transformed

> an immense tract of barren waste into a highly productive and ornamental country, enriched with abundant harvests, colonized with substantial and elegant residences, and above all, peopled with an intelligent, scientific, and grateful tenantry.[55]

This was a domesticated, arcadian vision according to the standard template, but with a progressive (if socially ordered) accent. Within such Whig agrarianism, the tenantry were to be scientific and grateful. The reverence for Coke and also his peculiar influence within the Whig party come here into sharper focus – he was the symbol of the Whigs' Janus-faced, deferential and yet enlightened conception. This ideal 'Coke' was no tory, was 'bigoted to no system' and prized 'no practice but for its utility'.[56] The sub-Baconian invocation in the *Hortus* of the need to save the appearances of nature by steadily amassing and recording facts was dramatised by its opposition (inevitably, Coke's opposition) to preconceived theories and unenlightened favoured practices. The productive harmony sought to integrate agriculture and manufactures; it was agrarian but not anti-industrial. Thus the research ventured into the best grasses for the manufacture of straw bonnets (under the aegis of the Duke's philanthropic and educational scheme for the children of the labouring poor at Woburn), finding, not surprisingly, that the straw recommended by William Cobbett's *Cottage Economy* 'proved too coarse for Leghorn plaits'.[57] With the perfection of its grasses, Britain would turn manufactures dependent on that material from articles of import to ones of export.[58] The argument of the *Hortus* was not protectionist, though it could by implication be made compatible with at least a moderate amount of protection. Instead, the experiments were presented as a defence of the agricultural interest (against the 'heavy national losses' that it 'has of late sustained') through the rational expansion of productivity.[59] This book, written under Bedford's patronage, presented a complex of ideas that was strongly anti-Malthusian, agrarian without being opposed to manufactures in an integrated landscape, utilitarian without being Benthamite and productionist without being necessarily protectionist. As such, it dovetailed closely with the political economy of Davy's own work, to which we now turn.

Aside from the pragmatic observation that it was intended to aid the feeding of an expanding population in time of global conflict, the political economy of Davy's chemical agricultural lectures has been overlooked.[60] Morris Berman assigns to Davy a somewhat passive role in the propagation of a set of landed class-based interests and ideology.[61] This social ideology vindicated social inequality by holding it up as the mainspring of agrarian progress and enlightenment. The praise and flattery of gentrified improvers, who were the governing strata at the Royal Institution, enabled Davy to get the necessary finance for his other projects, such as the construction of an ambitious voltaic pile (1809).[62]

However, when Davy proceeded from the premise that 'the connexion of Chemistry with Agriculture ... can hardly fail to be highly beneficial to the community',[63] he defined the benefit, not only by the improvement of subsistence and prosperity deriving from agricultural surpluses, but also by an expanding population and specialising workforce. Davy confidently declared that 'nothing is impossible to labour, aided by ingenuity'.[64] Britain's peculiar advantage in the use of machinery and division of labour could be applied to cultivation in the same way as to 'arms, commerce, letters and philosophy'.[65] Science was not merely the speculation of theorists but was 'the refinement of common sense, guided by experience, gradually substituting sound and rational principles, for vague popular prejudices'.[66] Scientific labour was materially expansionist, socially optimistic and historically progressive. According to Davy, 'as chemical philosophy advances towards perfection, it will afford new aids to agriculture', thereby harmonising the interests of classes.[67] Berman notes that the reconciliation of classes through enlightened progress was a common conceit of the period. Landholders, 'fitted by their education to form enlightened plans', would inevitably act as the conduits of improvement: 'in all cases the benefit is mutual;

for the interest of the tenantry must be always likewise the interest of the proprietors of the soil'.[68] But *pace* Berman, this was not only cheerleading for the 'ideology of science of the improving landlords'[69] – it was also the conclusion of a set of arguments relating to labour, population, incomes and exchange. In other words, it was not only a normative discourse concerning the impact of science on the relationship of classes (that is to say, a social ideology of science) but also a descriptive discourse about how economic processes worked and how they might be affected by physical processes. Naturally, this allowed such economic discourse to be shaped by agricultural chemistry.

Thus, Davy put forward a biomass argument relating to the balance of trade. In the eighth lecture, he contended that 'the exportation of grain from a country, unless some articles capable of becoming manure are introduced in compensation, must ultimately tend to exhaust the soil'. Fortunately

> in this Island, our commercial system at present has the effect of affording substances, which in their use and decomposition must enrich the land. Corn, sugar, tallow, oil, skins, furs, wine, silk, cotton & etc. are imported and fish are supplied from the sea. Amongst our numerous exports woollen, and linen, and leather goods, are almost the only substances which contain any nutritive materials derived from the soil.[70]

The present sterility of Sicily (he surmised) was probably due to 'the quantity of corn' carried off by the Romans. The desertification of North Africa and Asia Minor could also be traced to ecological deficits built up by places 'anciently fertile'.[71] Clearly, this was not a protectionist argument so far as the pivotal corn crop was concerned. Perishable importation was one type of intervention that would maximise the productivity of the soil. Nor was there much of a sense of Malthusian arithmetic frontiers:

> The soil offers inexhaustible resources, which, when properly appreciated and employed, must increase our wealth, our population, and our physical strength. Discoveries made in the cultivation of the earth are not merely for the time and country in which they are developed, but they may be considered as extending to future ages, and as ultimately tending to benefit the whole human race – as affording subsistence for generations yet to come – as multiplying life, and not only multiplying life, but likewise providing for its enjoyment.[72]

It was precisely this passage that was reproduced as the banner frontispiece of the *Hortus Gramineus*.[73] Production would provide not merely for subsistence, but for a degree of gratification. Also, there was a complementary relationship between the natural theology and the benign political influence of rational cultivation. If, on the one hand, 'the modifications of the soil and the application of manures are placed within the power of man, as if for the purpose of awakening his industry and of calling forth his powers',[74] it was also the case that 'the true objects of the agriculturalist are likewise those of the patriot'. By making cultivation more successful in the calling forth of labour, scientific agriculture made men 'love their country better, because they have seen it improved by their own talents and industry, and they identify with their interests, the existence of those institutions which have afforded them security, independence, and the multiplied enjoyments of civilised life'.[75] Bedford's patronage of Davy's programme pointed to the patriotism of national increase, a patriotism that underpinned the frequently conflicting patriotisms of national and imperial defence and of common liberties.

There were admittedly differences of emphasis between the Bedford and Davy models of patriotic increase. The most obvious lay in the social distribution of progressive knowledge. Davy's focus was on the gentry and aristocratic landlords, whose super-sentient leadership in the countryside would bring about improvement through the pragmatic obedience of their husbandmen and tenants. Davy was quite blunt about the social power that he hoped his lectures would endow his gentlemanly audience with. Whereas the common labourer 'can never be enlightened', he would follow methods that plainly worked, and moreover would not seek to deceive a master when he 'has conviction of the extent of his knowledge'.[76] In contrast, the patrician Whig model sought to maintain social hierarchy, but at the same time to ensure that the tenantry was also 'scientific' in its awareness.

It was this last consideration that was made an item of complaint in the *Edinburgh Review*'s treatment of Davy's published course of lectures, with data from the Bedford experiments appended.[77] The review made it clear that the public association between Bedford and Davy had been effectively signalled: 'In our own country, the name of Russell, so proudly distinguished in the annals of freedom, stands pre-eminent among those who have patronized this noble art.'[78] While praising the 'great care, intelligence and precision' of the Bedford appendix, the *Review* wished that it had been communicated less expensively, and thereby brought within the reach of the 'practical husbandman'. The appendix, with other matter, should be published separately and the remainder reduced to octavo in the second edition. The work would consequently 'fall in price, and rise in value; and thus would be more likely to pass into the hands of a numerous class of readers, who would assuredly return to their country, in the extended benefits of agriculture, the advantages they derived'.[79] While the lectures were republished in the smaller format, they retained all the material, and it was this second edition (of 1814) that was noticed by the *Quarterly Review*.[80] The latter piece acidly commented on the elevation of prices caused by the 'magnificent scale' of 'modern improvements', but acknowledged that some such inflation was unavoidable. For, 'If our population has increased, it is of urgent necessity that a greater supply of food should be procured for its consumption.'[81] Davy's lectures, with the Bedford connection, were recognised as interventions in political economy, as well as a corpus of chemical argument.

Despite their differences over the social distribution of knowledge, the conjunction between agrarian Foxite whiggery and Davy's toryism can be seen in their shared anti-Malthusianism, a refutation mounted by the patriotism of left and right. The discrepancy between Malthus' 'geometric' and 'arithmetical' ratios had destabilised the ideal arcadian harmony between population and subsistence. The patriotic riposte was to refuse to accept the abandonment of the harmonious model. Instead, it sought the expansion of production within the old balanced framework. The problem, in essence, was to put the georgics into motion. In one respect, this was achieved by abandoning traditionally static ideas of virtue. Davy put this argument forcefully in the moral overview of his introductory lecture on geology in 1805.[82] The conceit of natural man – his simplicity, pastoralism and unregulated freedom – was rejected in favour of the exertions of intelligence and invention.[83] According to Davy, the refinements of modern social intercourse were not the same as the luxuries of the ancients or of oriental empires, which were the passive gratifications of the senses. There was therefore no inbuilt mechanism of decline except wilful stagnation, the refusal to strive after excellence. Davy's aggressively modernist argument celebrated legitimately acquired wealth as the basis of credit and capital and their vital circulation like blood through the realm.[84] This did not mean, however, the diminution of agriculture by industry and commerce; on the contrary, the effort required to rear an enormous prize-winning Smithfield sheep improves the whole husbandry.[85] If the wants of man are subject to constant limitation by natural poverty – Davy's concession of the Malthusian premise – then the exertion of man will produce both subsistence and higher gratification through self-mastery.[86] Davy went on to make clear in his account of the formation of soils that the earth was made to respond to cultivation, the globe working through a larger equilibrium to support created life and happiness.[87]

In mobilising the georgic tradition, the Whig landed interest could draw upon its heritage of dynamic arguments concerning virtue. Whig rurality was not composed of unchanging simplicities. Whigs gibed at a caricature of tory squirearchy, not because of its agricultural mode of production but because it supposedly represented unreflecting idiocy (in a sense comparable to that later applied to the peasantry by Marx). In the 1790s the Whig concept of the independence of country gentlemen already involved 'improving' autonomy. The idea assumed (within an agrarian framework) the growth of primary commerce (for example, in agrarian products) within an exchange economy. Whig independence did not signify isolated self-sufficiency but possession of the resources for civil participation, both in deliberation with peers and of course in monetary transactions. It has been noted elsewhere that 'improvement' gained its force as a description of the transformation of techniques to foster agricultural productivity, and this was particularly true of Whig conceptions. For opposition Whigs, the potential independence of country gentlemen was not in doubt, though it was too often perverted by the machinations of the King's friends. The point here is the structure rather than the accuracy of the Whig narrative. Opposition Whiggism never really got to grips with the impact of the French Revolution on gentry sensibilities and the rise of a nationalist patriotism. Instead, Whigs diagnosed educational and intellectual problems among the country gentry. Ignorance and backwardness led to an absence of effective civil sociability. As Joseph Jekyll MP wrote in 1796, the country magistracy was short of both brains and liberality.[88] Jekyll wished that his patron Shelburne would read *The Case of Labourers in Husbandry* stated by one David Davies, a Berkshire rector.[89] Davies' urging of the cultivation of wastelands correlated with Lansdowne's interest in encouraging the diversification of wool-farming. Behind the smiling arcadianism of the ideal estate lay chronic concerns about poverty, population and trade in foodstuffs. The problem with Crown interference was not that it marshalled a country interest, but that it encouraged country partisans to be unthinking and selfish. Agrarian whiggery, in contrast, was both bucolic and rational.

As the periodical material indicates, both Bedfordian patronage and Davy's lectures on agricultural chemistry and geology were received in part as contributions to the ongoing debate over population. This modernised georgic or geoponic strand of argument serves to diversify our view of the ongoing ramifications of the controversy over Malthus. Recent productions in what Boyd Hilton has called the Malthus Industry have emphasised the role performed by the Malthusian incursion in breaking the optimism of *classical* political economy.[90] Within this 'moment' (essentially the first two decades of the nineteenth century) the significance of the essays on population has been reasserted. For Gertrude Himmelfarb, it is the lock-step of the Ricardian–Malthusian iron law of population that defines the turmoil of the age, rather than secondary disputes over theories of value, rent and glut.[91] Mary Poovey concentrates upon the relentless numeration in the architecture of Malthusian argument, presaged in the Essay of 1803, and consummated in the Essay of 1806.[92] Other historians recapitulate the relationship between the theodicy of the first essay and the sharper evangelicalism of Malthus' later writings.[93] All of these authors see Malthus addressing, revising and substantially (if politely) subverting Smithian expectations of long-lasting improvement through high wages, high consumption, the division of labour and the production of value through commerce and manufactures. Whereas Smith had criticised most varieties of corn law and tacitly approved of a system of poor relief, Malthus initially advocated protection to stimulate agrarian investment, and condemned the encouragement of surplus population through redistributive maintenance.

This view of Malthus as a kind of physiocratic fundamentalist – tightly relating the economic prospects of society to its fixed boundaries of subsistence – has tended to define itself in contradistinction to Smithian argument. Yet as we have seen, this was not the only context of reception. The Malthusian assumption of the arithmetic ratio was the weakest and least developed part of the theory, and was vulnerable to the political economy of agrarian productivity. Such a school of thought was not summoned into existence by the Malthusian challenge; instead, a long-established geoponic style transformed its character. Just as landscape genres in pictorial art adjusted and flourished, so this georgic tradition proved to be surprisingly resilient, reinventing its static aspect, and manifesting itself through natural science. This was a project undertaken by the aristocratic Foxite gentry, allied with scientific patriots of the right and left.

Its opposition to Malthus and protectionism helps to distinguish the Whig georgic position from other approaches. We have already seen how liberality was a competitive terrain, and the same was true of the georgics. Richard Drayton notices that the *Hortus* formed part of a Whig agrarian programme running parallel to Joseph Banks' regime.[94] In his treatment of the Lincolnshire squire and long-standing President of the Royal Society, John Gascoigne corroborates how scientific politics rooted in landowning could cohere with a vigorously imperial neo-mercantilist doctrine (in this agreeing with Bayly and Drayton).[95] Banks adopted a practical position similar to Malthus' in advocating agricultural protection to stimulate domestic production and innovation. At the same time he was more optimistic than Malthus about the capacity for human ingenuity to transcend natural constraints (here, Banks' approach resembled Davy's). Gascoigne points out that it was perfectly possible for Banks to function as an international man of science at the same time as pursuing the national interest as he saw it. Free trade in ideas did not necessitate free trade in commerce. Meanwhile, Banks was unsympathetic to Whiggery. He was a firm Pittite tory (describing himself as a tory supporting Pitt), a courtier – and agricultural adviser – to George III, and a confidant of Sandwich, Hawkesbury and Dundas. Banks rose to prominence during the 1770s, and it is worth noticing that in terms of political generations, Banks was a King's loyalist before this alignment became Pittite (after 1782–1783) and also before Smith became fashionable. It is therefore not surprising that Banks' anti-Smithian position should show some affinities with Malthus' later revision of Smith, while being more optimistic in tone and temper.

In his treatment of Banks Gascoigne has exposed a significant debate *within* Pittite circles regarding Smith's political economy. Banks' contemptuous dismissal of the Scottish 'monk' was directly related to his scientific expansionism. Colonial exploitation and domestic improvement enabled the natural cultivation of import substitutes. For example, one goal of the *Bounty* expedition (launched under Banks' patronage in 1787–1789) was to transplant breadfruit from the Pacific to the Caribbean in order to help to free the British West Indies from reliance on American food imports. Both Banks as an individual and the Royal Society under his influence acted as government consultants and were particularly influential at the Mint and Board of Trade. From 1797 Banks was a Privy Councillor. His role was more diverse and even more prominent than that played by Arthur Young *vis-à-vis* the Board of Agriculture. Gascoigne puts this in the wider context of British state-building, a state that was at war during the bulk of Banks' active intellectual life. Banks emerges as an active state-builder, comparable in this respect with the later civil servants touched by philosophic radicalism such as Chadwick. At the same time, Gascoigne emphasises Banks' self-professed identity as a country gentleman even after his activities became more metropolitan (and his mobility constrained by gout from 1800 onwards). This presented a powerful challenge to Whig models of georgic improvement; or perhaps more accurately, it represented a model that linked cultivation with executive power and administrative centralisation. Certainly Banks was closer to government than the Foxite opposition, and from this perspective we can see that Whig attempts to present a distinctive georgic argument were actually forays into territory owned by the Crown, loyalists and Pitt.

Georgics of the mind

The reception of the georgics was not confined to the estates of agrarian Whiggery. It also manifested itself in the thought and political economy of Edinburgh. Intellectual historians have not only underestimated the significance of Whig agrarianism outside the Scottish academy; they have also neglected the georgic character of the Whig side of the Scottish enlightenment. It was of course important for groups that were either opposed to or felt themselves to be distinct from Scottish 'Philosophic' Whigs to snatch away the legitimating ground of rurality. Yet the elaborate derision of Scotsmen inventing philosophic explanations for the misfiring of their shooting rifles, or the caricature of a Mac Quedy discoursing on 'smoke; steam, gas, and paper currency' in *Crotchet Castle* (1831), should not distract from the Virgilian frameworks that did exist in Edinburgh at the turn of the century.[96] The Broughamite Society for the Diffusion of Useful Knowledge (1826) took its agricultural duties as seriously as its industrial ones, issuing both the *Farmer's Series* and the triple-volumed *Manual of British Husbandry* (1834).[97] Much of this was due to Lord Althorp's influence, but Henry Brougham was eager to promote scientific agrarianism too (an impulse one can plausibly link to his readiness to play the part of the Westmorland gentleman).[98]

More subtly, interpretations of Scottish natural science (specifically, the chemistry and political economy of the earlier period c.1790–1820) tend to orientate them towards commerce and the prospective world of industry. Neither Fontana nor Chitnis touch upon agriculture, nor does Winch substantially discuss it.[99] According to Jacyna, chemistry was paradigmatic in Enlightened Edinburgh because it served the demands of a burgeoning ideology of industrial improvement and manufactures.[100] A partial corrective is provided by Maureen McNeil's account of the philosophies of agriculture that influenced Erasmus Darwin. These included the theories of Lord Kames, Professor Francis Home (1719–1813), George Fordyce (1736–1802), Dr Alexander Hunter

(1792–1809) and the Norfolk improver and uniformitarian geologist James Hutton (1726–1797).[101]

The predilection of young Whigs in the 1790s such as Francis Horner and Henry Brougham for visiting places of manufacture (of tobacco-pipes, for example, or steel) is well attested.[102] Yet a certain lack of precision in locating its intellectual moment has obscured our view of the significance of this interest. Chemistry proposed two distinguishable types of progress in production. The first corresponded to metaphors of growth or fertility, congruent with georgic political economy and the 'Agricultural Deism' identified by Dennis Dean in relation to Huttonian geology.[103] The second chemical model invoked material addition and serial expansion through the building up or recombination of elements, understood in various ways. The key break between them came not with the 'chemical revolution' of Lavoisier and Black, but with the Daltonian atomism that rose to salience from 1808 onwards. It is important to remember that the intellectual formation of the original Edinburgh Reviewers occurred before the Daltonian break. The chemical world of Whig statesmen such as Horner and Brougham was recognisably the world of the earlier Philosophic Chemistry of William Cullen (1710–1790) and Joseph Black (1728–1799). This was a study directed towards topics such as the potency of manures, framed by concepts of respiration and combustion and the doctrine of latent heat. Such chemistry was oriented towards similar practical problems to those that exercised the agrarian English Foxites, namely enlightened increase in agricultural productivity in order to maintain a harmonious social order. The similarities suggest that certain georgic perspectives were shared between some Edinburgh intellectuals and agrarian Foxites. The outlook was not identical, but there was enough common ground for political signals to be sent and understood.

The distinct ramifications of the Scottish georgic style were shown in specific debates over Virgil, in georgic political economy and in the invocation of the Virgilian–Baconian 'Georgics of the Mind'. As an authority, the didactic Virgil enjoyed a somewhat ambivalent status in the Scottish academy, partly because he was seen as an important though difficult author who was easy to misunderstand. One manifestation of this sensitivity was anxiety about the possible contamination or manipulation of Virgil's meaning in translation. That was probably why Professor Dugald Stewart always quoted Virgil in the original Latin in the body of his text, and frequently provided his own English versions in footnotes. Stewart also manifested coolness towards Dryden's famous translation, criticising, for instance, his rendering of the relationship between mother and child in the fourth *Eclogue*. By obscuring the meaning of the word 'incipe' ('begin'), Stewart charged that Dryden had diverted the reader's attention away from the instinctive comprehension of natural signs.[104] The boy 'began' to know his mother by her smile, because of the human propensity to learn by the repeated connection of inherent passions and effects. Elsewhere in the second volume of *Elements of the Philosophy of the Human Mind* (1814) Stewart repeated the same point, again criticising Dryden for mistranslation, and stating that the associative psychological theories of David Hartley and Joseph Priestley could not account for the 'pathetic effect' of this natural sign.[105] There is an echo here of the Reidian distrust of English associationism. It involved, in this case, an aesthetic re-appropriation of Virgil's verse. Unease over Dryden's translation, however, was not new, and was a feature of writers modelling themselves on the earlier 'Geoponic' genre. In the first of his *Essays on Husbandry* (1764), the Reverend Walter Harte opined that if Dryden had submitted his performance to an Italian husbandman or 'plain English yeoman', instead of referring himself to professed wits and critics, 'the poetry of our English Georgics might have been a masterpiece of truth, as well as fine writing'.[106] Harte was one of Stewart's sources, and we shall return to him below. The point to grasp here is that a similar distrust of Dryden's diction could arise from different intellectual preoccupations, including a concern for the didactic content of Virgil's verse.

In the context of natural science Virgil was considered first of all as a conduit or exemplar of method, against whom the superiority of modern enquiry could be delineated. According to Adam Smith's twenty-fourth lecture 'On Rhetoric and Belles-Lettres', delivered at Glasgow in 1763, Virgil's *Georgics* comprised an inferior species of didactic eloquence.[107] This was in fact an Aristotelian mode of argument that could be directly compared with the Newtonian method, prefigured by Descartes. Virgil's procedure had been to advance a different principle for every phenomenon to be explained. He had tried to account for various types of cultivation separately: in the first Georgic, corn; in the second, trees; in the third, cattle; and in the fourth, bees. If, however, Virgil had begun with the principle of vegetation and then proceeded to the appropriate culture of particular plants then this would have approximated to the Newtonian and Cartesian methods of didactic system. By this more philosophic procedure, according to Smith, one could reach a multiplicity of rules or phenomena on the basis of the fewest principles necessary and in a properly ordered manner.[108]

Smith's disparagement of Virgil's method in 1763 was essentially reproduced by Dugald Stewart in the *Elements of the Philosophy of the*

Human Mind (1814). According to Stewart, Virgil was guilty of confusing laws with facts. In the first *Georgic*, the poet had placed different fruits 'in stated regions, from the eternal cause, / Such nature's compact and unbroken laws'. In the fourth *Georgic*, on the bees, Virgil had written that 'They, sole, their town, their race in common rear, / Know their fix'd households, and just laws revere.'[109] But by describing as laws 'those physical arrangements, whether on the earth or in the heavens, which continue to exhibit the same appearance from age to age', Virgil had suggested that laws operated as efficient causes.[110] According to Stewart, however, laws were the generalisation of 'established conjunctions of successive events'.[111] To set up a law for every constantly acting cause would be to deny the ladder of generalisation, first recognised by Sir Francis Bacon and subsequently given mathematical extension by Sir Isaac Newton.

Worse still, the self-containment of facts as laws opened the way to the 'extravagant hypothesis' of cyclical time.[112] In the prophetic fourth *Eclogue*, Virgil had intimated the alignment of moral with astronomical cycles: with the rise once again of a new Troy.[113] Stewart likened this to the belief, given some credence in Bacon's essay 'On The Vicissitude of Things', that at the end of the *annus magnus*, or Platonic Year, 'a repetition would commence of all the transactions that have occurred on the theatre of the world'.[114] Stewart cautioned against a contemporary vogue for this kind of cyclical argument, precipitated by Joseph Lagrange's *Traité de mécanique analytique* (1788), which had been reissued in the first volume of its second edition in 1811.[115] Lagrange had established, using an algebraic extension of Newtonian mathematics (although against Newton's conclusions), that irregularities in planetary orbits were subject to periodic laws, and consequently that the celestial system did not contain the elements of its own decay. Stewart reserved comment on the implications of this discovery, emphasising instead that it did not constitute proof of the cycles of antiquity.

This criticism of Virgilian notions of time helps to put into context an observation made about Edmund Burke by J.C.D. Clark. Clark observes in his *Reflections on the Revolution in France* (November 1790) that Burke's account of historical evolution looked back to Virgil rather than forward to Darwin.[116] Clark has in mind Burke's invocation of the beehive in the fourth *Georgic*, in which a single community might subsist, transcending the short lives of its members. Burke's view is not cyclical in a strong sense, but the oscillation between conservation and correction is aimed at preserving an immortal constitution. Burke's conception was of a static macroscopic harmony, within which there could be self-adjusting internal motion. But by rejecting the idea that a specific phenomenon such as Virgil's hive could constitute its own law, Stewart was substituting an anti-cyclical progressive view of phenomena in time, subject to laws of ever-higher generality. Now, this progressivism is hardly surprising. What is worth noticing is that different emphases in the temporal interpretation of Virgil provided a measure of the theoretical difference between Burkean Whiggery and Scottish Philosophic Whiggery. This was a tension not between an organic theory and a mechanical theory, but between a static and an open description of mechanism in historical nature.

Ambivalence about the didactic Virgil did not, however, prevent the dissemination of georgic influences in the Edinburgh intellectual world, as is revealed when we consider which authors were read alongside Malthus by a generation of Scottish university students. On 19 January 1802, at the request of his fellow Edinburgh student John Seymour (1777–1819), Francis Horner drew up a reading list on political economy for Seymour's elder brother, the Duke of Somerset.[117] Horner's sources were Stewart's lecture recommendations supplemented by his own library discoveries.[118] Horner divided his reading list into two sections: population and money. Included under the former head were Arthur Young's *Travels in France* and *Tour in Ireland*, 'The Statistical Works' of Sir William Petty (1623–1687) and Charles Davenant (1656–1714), Benjamin Franklin's essays and Robert Wallace's *A Dissertation on the Numbers of Mankind* (1753). *The Essay on the Principle of Population* by Robert Malthus (1798) merited a brief entry with no explication; Horner appeared unfamiliar with it, noting the author's name incorrectly.[119] This note corroborates Poovey's observation that the impact of Malthus' *Essay* was not fully felt until the second and third editions of 1803 and 1806.[120] More prominent in Horner's list were the Reverend Walter Harte's *Two Essays on Husbandry* (1764). Although now obscure, the essays ran to a second expanded edition in 1770 and were referred to by Adam Smith in the *Wealth of Nations*. Horner especially recommended it, commenting that Chesterfield had praised its style in one of his letters.[121] The letter in question, of 3 September 1764, had compared Harte's prose to Virgil's verse.[122] According to Horner, the argument advanced by Harte was quoted approvingly by Dugald Stewart, particularly in criticism of Colbert's policies. The treatise placed a flourishing agriculture at the root of the expansion and social happiness of communities. The second of Harte's essays comprised an account of experiments to increase the growing of lucerne (alfalfa), a key fodder plant. It is therefore no surprise to find the essays listed by Horner alongside the writings of the Physiocrats, principally the *Philosophie Rurale* of the Marquis de Mirabeau, which Horner considered the most systematic and elaborate statement of the Physiocratic doctrines.

There is, to put it mildly, a sharp discontinuity between Harte's pastoral and the Malthusian pressure on subsistence. Malthus was relatively uninterested in the agrarian idylls of Greece or pre-Augustan Rome. In the historical third chapter of the *Essay*, the savage or hunter state is embodied in the North American Indians or Cape Hottentots, while the shepherd state is represented by the barbarian tribes that overran the Roman Empire.[123] According to Malthus, the Scythian shepherds were driven forward like wolves by want: a gathering body of darkness and terror that sank Italy and the world into night.[124] Savage wars of extermination thereafter raged in the constant struggle for room and food. In a dark analogy with the Fall of Man, Malthus conceived of younger sons issuing forth to engage in the struggle for existence. Like Adam and Eve in *Paradise Lost*, the world lay before them to choose.[125]

The religious and classical temper of Harte's *Essays on Husbandry*, a text some thirty years older than Malthus' work, is very different. Born in 1709, Harte was a disciple and frank imitator of Alexander Pope, carrying forward Pope's mixture of moderate reason and conservative natural theology, his 'Cosmic Toryism'. The Fall of Man did not only bring pains, but the possibility of temporal reward. Industry was God's appointed means of alleviating his malediction: 'all meliorations and improvements, in the culture of the earth, are divine rewards, proposed and reserved for man, as the retribution of his diligence'.[126] As God's punishments and rewards were universal, and diligence was to be encouraged, 'it is certain that every soil is capable of being improved'.[127] For, as St Paul had proclaimed in Corinthians, 'we are all God's husbandry', and 'the georgical history of Boaz, Ruth, Naomi and Orpah' had united worship with cultivation.[128] The 'Georgics of the Mind, as Lord Bacon expresses himself', including the essays themselves, therefore expressed rational devotion. Meanwhile, in Harte's estimation, 'Rome was ruined more by neglect of agriculture, and giving no attention to useful trade and commerce, than by the invasion of barbarians.'[129] So far as the course of cultivation was concerned, the collapse of Rome did not constitute a major interruption. Agriculture 'continued in a sort of declining condition from the days of Virgil and Columella till the time of Constantine IV, and then lay in a kind of dormant state till about the middle of the reign of Henry VIII'.[130]

In Stewart's *Lectures on Political Economy*, Harte is particularly referred to for his critique of Colbert,[131] and it is this (as well as the general recommendation) that was passed on by Horner. Harte condemned Colbert for having encouraged the over-production of *clinquallerie* (that is, jewellery, clocks, watches, lace and embroideries). In so doing he had 'half-ruined husbandry in France, by making an attempt to disinherit agriculture, and adopt trade; whereas in truth, he ought to have encouraged both, and caused them to flourish at the same time'.[132] Harte argued so that 'the two occupations may be justly harmonized, but that the scale may preponderate a little in favour of husbandry'. Equilibrium guaranteed the flourishing of the whole; husbandry in this broad sense did not entail an exclusively agricultural economy.

Texts like Harte's were helpful in putting the georgics into motion. Yet the most effective way to put Virgil to work in the service of a forward-moving history was to place him under the penumbra of Sir Francis Bacon. As we have seen, Virgil was held to epitomise approaches to natural philosophy that had been superseded. The critique of Virgil was seen as part of the story of the forging of more modern conceptions. But the reasons that made georgic arguments obsolete in natural science helped to render transformed uses respectable in the realms of morals, economics and politics. The effect was reflected in Sir Francis Bacon's revision, presented in *The Advancement of Learning* of the 'georgics of the mind'.[133] The conjunction of Virgil and Bacon, cultivation and science, was disseminated by Stewart, the clearest statement being his 'Preliminary Dissertation' to the *Encyclopaedia Britannica*. This was first issued in the Supplement of 1815–1824, and was published in a revised form in 1842.[134]

In both versions, the Baconian legislator was praised for his eudaemonistic intention, expressed in *De Fontibus Juris*. This was 'that the citizens may live happily' through well-defended and policed education, moral instruction, loyalty to magistrates and material expansion (the citizenry 'should abound in wealth and other natural resources').[135] Stewart pointed out the correspondence between these aims and 'Mr. Smith's ideas on the same subject', expressed at the end of *The Theory of Moral Sentiments*.[136] A vital problem upon which Stewart brought Bacon to bear was that of how the legislator might be considered to be productive or creative. The education of the people, in particular, allowed the legislator 'the abundant harvest which rewards the diligent husbandman for the toils of spring'.[137] According to Stewart, by bestowing on education the title of the 'Georgics of the Mind', Bacon had combined

> the two proudest functions entrusted to the legislator – the encouragement of agricultural industry and the care of national instruction.

> In both instances, the legislator exerts a power which is literally productive or creative; compelling, in the one case, the unprofitable desert to pour forth its latent riches; and in the other, vivifying the dormant seeds of genius and virtue, and redeeming, from the neglected wastes of human intellect, a new and unexpected accession to the common inheritance of mankind.[138]

The passage is worth citing in full because it is not original, that is not original to the 'Preliminary Dissertation'. Instead, Dugald Stewart lifted it verbatim from his own *Lectures on Political Economy*, first delivered in the winter of 1800–1801 and thereafter every year till 1809.[139] The original context of the georgic argument was Stewart's defence against Smith's view that authors merely 'reproduce annually the value of their own annual consumption'. The problem was extensible to all those who produced perishable performances with intangible effects, such as musicians. It was most acute, however, in the case of philosophers and statesmen. Stewart insisted that 'the labour which is employed in the cultivation of the understanding approaches more nearly (in the harvest which it yields) than anything else which can be specified, to the labour of the husbandman'.[140] The parallel was literal in that both types of cultivation involved the moulding of the 'bounty of nature', both moral and material. While disagreeing with their terminology, Stewart partially defended the distinction between agricultural and manufacturing labour drawn by the Physiocrats: not that the one was productive and the other was not, but that the direct combination of human industry with the bounty of nature was a special sort of production. Philosophers and statesmen may not have been manufacturers, but they were cultivators. Consequently, the legislator, in his promotion of the georgics of the mind, was productive in the terms that were repeated in the 'Preliminary Dissertation'.

Returning to this text, Stewart implied that Bacon's georgics of the mind did not only or narrowly refer to scholastic instruction. The cultivation of the understanding and the improvement of expression that was both its means and its effect enabled the formulation of a Baconian science of *Leges Legum*: a generative grammar of laws.[141] There were two forms of this science. The first was an etymological and historical study of leading and governing rules, expressing the nature of a society, and the *Leges Legum* proper, 'a theoretical model of legislation . . . a standard for judging the comparative excellence of municipal codes which may suggest hints for their correction and improvement'.[142] One was diagnostic and interpretive, the other prescriptive. They differed from each other 'not less widely than the rules of Latin or Greek differ from the principles of universal grammar'. But each, especially the latter, was the fundamental labour of the scientific statesman.

The prescriptive science of laws, of precise rules for discriminating between codes and thereby reforming them, was derived from the inductive study and classification of social and legal phenomena. It was a matter of experience, and the accurate description of experience. Stewart quoted with approval the aesthetic praise lavished on Bacon by Charles James Fox.[143] The Lord Chancellor and Raleigh and Hooker were ornaments of intellectual cultivation in those happy years of peace between 1588 and 1640, an era, as Stewart pointed out, that witnessed the growing Jacobean demand for books on agriculture.[144] This was a very selective reading of Fox's unfinished *History*, which was published posthumously in 1808. According to Stewart, the reform of style under Elizabeth and the early Stuarts, as described by Fox, comprised an important instance of the intellectual georgics. As the chief source of error was the imperfection of words and language was the means of spreading enlightenment, 'whatever tends to diminish the ambiguities of speech, or to fix, with more logical precision, the import of general terms' was essential to the progress of mind. Command over expression, cultivated in this manner, enabled the business of legislation – analogy, classification, improvement – to be articulated and pressed forward.

The 'georgics of the mind' constituted a critical revision, in the *Advancement of Learning*, of Virgil's vocabulary. Bacon's point, in the section of the second book dealing with the appetite and will of man, was that Virgil had gained as much glory from his account of pastoral toil as he had from the heroics of the *Aeneid*, his epic on the foundation of Rome: 'And surely, if the purpose be in good earnest ... really to instruct and suborn action and active life, these Georgics of the mind, concerning the husbandry and tillage thereof, are no less worthy than the heroical descriptions of virtue, duty, and felicity.'[145] The broader polemic was that Virgil had drawn a false distinction in the *Aeneid* between the sciences of peace, attributed to the Greeks, and the imperial arts of conquering Rome. A similar contrast was picked up by Stewart and his pupils. Through the Baconian revision, Edinburgh Whiggery had taken the pastoral, cultivating and refining aspects of arcadia, but had harnessed them to a moderately expansionist and ameliorating sensibility. This meant jettisoning the temporal equilibrium and static harmony of the original. By tending the Georgics of the Mind, an indefinite improvement in social and intellectual conditions was possible.

Yet despite the inventiveness of the tradition, and the range of styles that it comprised, the georgic register was waning by the middle of the nineteenth century among both rural-oriented aristocrats and Scottish academic Whiggery. Virgilian discourse proved to be less hardy than the classical republican canon, which, as Eugenio Biagini has argued, remained tenacious within Liberal politics into the 1870s.[146] The georgic re-orientation, and the counteroffensive of the first two decades of the nineteenth century, abated in the period before the recession of landed wealth in the 1870s. The georgic style was comprehensive, harmonious and patriotic, being rooted in a sense of landed location and an understanding of cultivation that took the intimate connection of place and fertility for granted. Its scientific liberality was not dependent, in the first place, on conceptions of the market, the state or even civic virtue, although it had implications for all of these political coordinates. In the Baconian moral and political vision of the 'georgics of the mind', the statesman adopted an analogous pose of philosophic cultivation, tilling pastures of political economy, education and legislation. The wider georgic vision of a balanced and harmonious bucolic economy, put into motion by the patriotism of national increase, retreated before more compelling styles of analytic and recombinatorial replication. Another agent of dissolution was the aura of crisis around the late Malthusianism of the 1840s, which, when it receded, left an industrialised conception of progress to take advantage of the renewed optimism of the post-Crimean decades. Thus, the georgic style was undermined by an analytic shift in chemical theory and politics, which had the cumulative effect of dissociating cultivation from immemorial environments. Politically, the generation of Foxite agrarianism yielded to a centrist Whig–Peelite formation. Tracing this trajectory is the task of the subsequent section.

Georgic decline: Whigs and Peelites

The replacement of the patriotic nexus of the Napoleonic period by a Whig–Liberal analytic style was exemplified by the cooperation of landed Whigs and Peelites in the Royal Agricultural Society (1838), the Royal College of Agriculture at Cirencester and the Royal College of Chemistry at London (1845). The aim of describing this coalition, particularly in the context of the College of Chemistry, is not to replicate the institutional history furnished by Gerrylynn Roberts, but to trace the shift in political style away from scientific liberality, and the georgic tradition, and towards analytic, free-trade liberalism.[147]

The generational change in Whig agrarianism had been signalled by the near collapse of the Smithfield Club in 1821. In that year Bedford withdrew from the presidency on the grounds that its objectives (centring on the fattening of stock) had largely been achieved. There was then a hiatus of four years before Lord Althorp took over the presidential position.[148] Like Bedford and Fitzwilliam, Althorp was anti-Malthusian, correlating population with increasing national prosperity. However, he was a discreet supporter of corn-law repeal and Poor Law Reform, so the change in leadership witnessed a definite liberalising shift in economics.[149] Goddard points out that it was Althorp, as the third Earl Spencer, who proposed the formation of the English Agricultural Society (shortly the Royal Agricultural Society of England, with the motto 'Practice with Science') at the Smithfield club dinner, in December 1838. Wasson has described Althorp's pivotal role in the establishment of the Agricultural Society in some detail, but marks off his scientific agriculture from his Whig politics, referring it instead to the aristocratic social leadership of technical change.[150] But this depoliticisation of agrarian improvement occludes the coalitionist nature of the change, and the transformation of style.

Thus the list of 'Noblemen and Gentlemen' on the provisional council of the prospective college of chemistry, forwarded to the Prime Minister Sir Robert Peel in 1844 (by the Whig Lord Ducie), shows a marked bias towards liberal Whigs and Peelites supportive of free trade.[151] Apart from the pragmatic protectionist Duke of Wellington, the MPs and Peers divided fairly equally between Whig and Peelite free traders, with some weight on the Whig side. The evangelical peer who presented the prospectus, the second Earl Ducie (1802–1853), came from the Whig Moreton family and had a high profile as a public advocate of corn-law repeal. The Lansdowne interest was represented by the third Marquess himself and his protégé Lord Monteagle (1790–1866), the former Chancellor of the Exchequer Thomas Spring-Rice. Other liberal Whigs included the enthusiastic cattle-breeder Lord Portman (1799–1888), Lord Clarendon (1800–1870) and Althorp's old ally Viscount Milton, now the third Earl Fitzwilliam (1786–1857). They were accompanied by the third Earl Radnor (1779–1869), the former Foxite and corn- law moderate, who had opposed the law of 1815 but supported the compromise of 1828. A more zealous corn law repealer and reformer was Sir Thomas Wyse, Catholic MP for Waterford (1791–1862). Of the Leveson–Gower connection, the Duke of Sutherland was temperately inclined towards the Whig party, his wife having been mistress of the robes during the bedchamber crisis. His brother Francis Egerton, MP

(1800–1857), had started his political life as a fervent Canningite and was a strong repealer. On the Peelite side this group was joined by Philip Pusey (1799–1855), the MP for Berkshire and ally of Gladstone, and by Earl Dalhousie (1812–1860), Peel's current Vice-President of the Board of Trade. In all, the cast of dignitaries represented a considerable portion of the centrist, liberal wings of both the Whig and the Peelite parties.

This patronage cluster of the mid-1840s contrasts suggestively with the improving cliques of the Napoleonic period. The Foxite generation had passed on; there were no equivalents to Charles Western or Thomas Coke. Compared to the political diversity of landed groups attached to the Board of Agriculture and its milieu at the turn of the century, the formation was ideologically more compact, at least so far as economic issues (in contradistinction to issues of political reform) were concerned. The interest of this particular cross-party formation lies in the common platform, a set of premises that both Whigs and Peelites were able to subscribe to (and that were therefore implicitly consistent with the diverse elements of their ideological identities). To be more precise, this was a grouping that was able to articulate a broad measure of consensus on the infrastructure (that is, the structures of education, training and information) that was to underpin agricultural trade liberalisation. The decade between the mid-1830s and the mid-1840s witnessed the coincidence of liberalising ideologies, which from very different premises coalesced around similar policy prescriptions. There is a sharp contrast here with the ideological polarisation of the immediate post-Napoleonic period. The existence of this common platform is an early indication of the passing of a period regarded by intellectual historians as a time when elite culture was fissured by polar opposites.[152] The platform can be taken as anticipating the more amiable public discussion of the subsequent third quarter of the century (which included a more conscious subdivision of intellectual labour).[153] In the first instance, this can be seen as a successful application of Whig manners of coalition building. However, the consequence was eventually to undermine the ethos of liberality and the georgic tradition.

With this in mind, it is useful to revisit the joint *Prospectus* in order to analyse its character as a public signal. Roberts points out that the promoters of what became the Royal College of Chemistry (the apothecary John Gardner and druggist John Lloyd Bullock) were ambitious entrepreneurs, who initially tried to establish a 'Practical Chemistry School' attached to the Royal Institution.[154] Although the attempt was unsuccessful, the nucleus of their support largely grew out of the membership of this body.[155] It is noticeable that Thomas Wyse was the principal political connection of the provisional committee mentioned by Roberts as working towards a chemical school from 1842 onwards.[156] Wyse figures again on the 'Prospectus' list of 1844. The list of 'Noblemen and Gentlemen' unanimously voted to vice-presidencies of the Royal College of Chemistry in August 1846 shows a substantial overlap with the 'Prospectus' list, including Lansdowne, Essex, Fitzwilliam, Clarendon, Ducie, Portman and a set of other prominent Whigs (Russell, Palmerston and Brougham).[157] A Peelite addition to the latter list was Sidney Herbert, MP. The overlap encourages us to treat with some caution Roberts' contention that those named on the prospectus were not always aware that they had been included: some being cited merely for expressing an interest.[158] Roberts' source for this is Jacob Bell's censorious editorialising in the *Pharmaceutical Journal*, yet the context here was Bell's ongoing vendetta against Gardner (acknowledged by Roberts) over the credentials required for Pharmaceutical Society membership, and also his discontent with the disrespect allegedly displayed towards British chemistry by Gardner.[159] The commentary makes it clear that Bell's anecdotage was directed against the projectors' motives and framed his expression of satisfaction that Gardner was no longer Secretary of the society.[160]

On the other hand, Bell's rumour raises the question of what was actually signified in scientific patronage. Bell wrote up a reported conversation in which 'A' and 'B' (who had had their names listed) ran into one another and found that each had allowed his name to stand on the patronage list for the new College of Chemistry because he thought that the other had consented to do so.[161] Unsurprisingly, there is no indication as to the identities of 'A' and 'B' and therefore little way of checking whether the episode was sharp practice on the part of Gardner and Bullock, or malicious gossip on the part of Bell. Moreover, Bell's argument was that this supposed deceit was unnecessary. The College overcame its initial problems, won spirited noble and professional support and thereby gained strength and influence.[162]

In the light of the Whig–Peelite overlap in the prospectus and the vice-presidency lists, the character of the prospectus as indicative of an intellectual rapprochement (even as a canny appeal by Gardner and Bullock to shared sensibilities) is tenable, and its argument will shortly be analysed in these terms. However, it is necessary to consider the alternative case: that is, that Bell's rumour was essentially true, at least for a couple of the notables. The structure of Bell's aspersion is also worth contemplating, whatever happened at the actual encounter that might have lain behind it. Clearly, the condemnation was of the projectors, not the sponsors. There was no opprobrium attached to notables who put their names forward in order to align themselves publicly with those with whom they wish to be associated. The implicit accusation was that Gardner and Bullock exploited this convention. Their conduct, now revealed, would astonish the aristocratic patrons.[163] The upshot of this observation is that however superficial the patrons' understanding, this was a group of statesmen who wished to be named together in the context of scientific improvement. As evidence of a Whig–Peelite, centrist conjunction, the list still stands.

Gardner and Bullock were plausible individuals – 'moral adventurers', rather than 'moral entrepreneurs'[164] – who were able to attract distinct coteries and through them to exploit established resource networks. Roberts recognises their financial entrepreneurship in their rather maladroit attempts to profit from various fixes (manoeuvres that confirmed the antagonism of Bell), but it was the intellectual or ideological manoeuvre that should be emphasised in the present context.[165] Roberts conceives their initiative as an appeal to a series of socio-economic constituencies (such as the land, medicine or industry), whereas the need to fashion a coalition between political actors was just as operative. The platform can be read as evidence of the ideological nature of Whig–Peelite cooperation, or rather of the grounds upon which a common platform could be pitched to both wings of moderate liberalism. The significant features of the *Prospectus* that need to be elaborated in the context of the present argument focus on this political intersection rather than on the institutional history of interests.

The vision of the chemical and social argument was both analytic and diffusionist. As Roberts recognises, the need for a specialised College of Chemistry was expressed in terms of the national interest and of competition from progressive Continental scientific powers.[166] What is crucial is the social and ideological mechanism identified as characteristic of the specialised German model which Gardner and Bullock sought to realise in their proposed chemical establishment. The model notionally worked to achieve its goals by bringing together, instructing and mobilising comparable groups of relatively modest carriers of knowledge: sons of manufacturers, the 'pharmaciens of every town and village [subsequently] qualified to perform analyses, and advise the farmers and smaller manufacturers of their neighbourhoods', and especially 'that numerous class of gentlemen, who, possessed of a small independence, are anxious to find employment which must with ordinary diligence lead to emolument and honour'.[167] In this respect, the model bears a closer resemblance to the older Whig preference for progress through the wider diffusion of knowledge (exemplified in Bedford's *Hortus*) than to the sharply hierarchical model preferred by Davy. Yet in addition there was the socially analytic character of the mechanism – men from diverse constituencies were capable of being impressed with correct curricula, and moulded into a comparable cohort that would then freely act to promote national productivity. The carriers of knowledge were to be drawn into the institution from various social backgrounds, instructed together and then sent into circulation: 'Every town and village ought to possess a person capable of analysing soils, animal and vegetable productions, and of teaching the principles of the Science.'[168] There is a parallel here with the analytic hopes invested in the application of the subject itself. Leaning heavily on Liebig, the *Prospectus* declared its confidence that 'a key has been furnished us to unlock the mysteries of organised beings'.[169] This was through resolution of the 'elements of which plants and animals are composed'. Wealth was realised through analysis: the Liebigian separation of sulphuric acid and the analogous possibilities of breaking down the soil so as 'to render the materials existing in it available to plants'.[170] Both the social form and the form of the science it carried rested on a conception of analytic dissemination. In other words, both the society and the science were atomistic in character, relying on the circulation of the proper personal and chemical elements.

This scientific model was therefore inimical to the unified and comprehensive sensibility of the older scientific liberality. On the other hand, it *was* liberal in the more modern, reductionist, individualist sense. It proposed an institution whose function would be to underpin a self-acting social mechanism of cognitive diffusion and improvement. It thereby attempted to broker a concordance between core magisterial Whig and Peelite ideological concerns. It was progressive and ameliorative, with a Baconian nod towards the augmentation of human power and happiness.[171] It was minimally paternalist or dirigiste. Chemical education might be established by aristocratic patronage and public authority. Nevertheless, it would serve to generate a social process of diffusion. It would thereby substitute for other forms of interventionism:

> The great change which has recently been made in the laws which regulate the importation of grain render it highly expedient that Landowners and Agriculturalists should seek some means of successfully competing with the foreign corn grower, and to the Science of Chemistry, they may confidently look for the desired assistance.[172]

Chemistry would assist by marshalling cognitive power, but it would not guarantee profits in the same way as legislative protection. The way was clear for producers to receive their just deserts according to their assiduity in applying the new analytic knowledge. While the exercise could be read in an expansionist light (most congenial to Whig assumptions) it was also restorative in character, defending the equilibrium of home production through the dissemination of chemical information. This was more of a Conservative, Peelite move. It is noticeable that the argument from improved productivity was primarily defensive, and oriented towards maintaining home growers' share of the domestic market.

It is not quite correct to state, as Roberts does, that the application of chemistry to agriculture was equally consistent with protectionist and free-trade arguments (thereby stepping outside this line of trade confrontation). While tory paternalists such as the Duke of Richmond were prepared to support scientific agriculture in the Royal Agricultural Society, the principal patronage market for schemes such as the College of Chemistry came from free-trading Whigs and Peelites. It was this political constituency that was appealed to by moral adventurers such as Gardner and Bullock. Of course the appeal could be widened as a consequence of both virtue and necessity. Whether presented as *faute de mieux* in response to tariff reduction or as the enabler of liberalisation, the argument for scientific agriculture was couched in terms of superseding protection. There was a well-worn Peelite register of response to 'inevitable' change (honed through Catholic Emancipation and the Reform crises) that could be applied to agricultural chemistry in this context.

Despite Gardner's and Bullock's success in yoking a number of prominent Peelites to the cause, their initiative ultimately failed to gain the backing of the most important Conservative, Sir Robert Peel himself. It is necessary to appreciate, however, that Peel turned down the proposal for a separate, additional college of chemistry because he considered it to be redundant. The function of dissemination was already being accomplished by existing institutions. The problem was not so much that the proposed college would have required an outlay by the state. After all, as Peel pointed out to Liebig (in reply to a suggestion that the Crown, aided by his laboratory, might exploit coprolite deposits), public funds were already directed towards agricultural chemistry.[173] The projectors made a strategic mistake in approaching Peel, presumably because they misjudged which pattern of diffusion he would favour. In his reply to the Whig Lord Ducie, Peel maintained that the study of chemistry would be better combined with that of other subjects in the universities.[174] Now this was directly contrary to the research specialisation that the projectors quite reasonably presented as a Liebigian approach. Liebig's *Familiar Letters on Chemistry, and its Relation to Commerce, Physiology and Agriculture,* together with the Giessen model, had been a mainstay of the *Prospectus*. Yet Peel presented his negative assessment as a report of Liebig's opinion, the Prime Minister having met the chemist on the previous day.[175] Paradoxically, this meant that Peel showed greater old-fashioned liberality (in this instance resistance to atomistic specialisation) than those of his followers who were cooperating with moderate Whigs in pushing an atomistic model. This hints that liberality was becoming less distinctively Whig at the same time as it was being superseded. The process was untidy, but the former Whig attachment to the older liberality was becoming attenuated as a new analytic Liberalism emerged.

Conclusion

Whig statesmen and science

As stated at the beginning, this study is not a history of scientific institutions or of practitioners within particular disciplines. It is a review of political manners: of ways of being a Whig statesman. The essay has taken a critical look at the changing aspiration of 'statesmanship', which consolidated its status by the turn of the nineteenth century. No pretence at comprehensiveness has been made: instead, a series of suggestive episodes have been presented to give an impressionistic view. Ideological trends and modes of display followed a *rough* chronology but not necessarily a continuous one: styles could re-emerge 'out of time' in response to certain events (Lord John Russell's response to the Ecclesiastical Titles controversy is an example). This pointillist approach is useful mainly for suggesting lines of future exploration, though some conclusions can tentatively be drawn. The study discovers that the Whig encounter with natural knowledge was intended to capture a certain kind of liberality. In one manifestation it expressed the Whig interpretation of the georgic tradition. Such strands are to be understood as modes of public posture, or style, rather than as ideologies or purely theoretical phenomena. Whig polemic assumed a scientific liberal character when it advanced the generous and therefore emancipatory effect of certain kinds of cognition. Historians of science have been rather too respectful of the methodological integrity of some of these strategies. At times crude and transparent and at other times manifesting subtle ideological bias, Whig conscriptions of science were pretty tendentious.

The georgic tradition was drawn upon in influential poses of cultivation. It comprised a lineage of harmonious, pastoral ideas that were inherited by Whig politicians. This canon, developed from Virgilian sources, bears comparison with the more familiar tradition of civic republicanism. Eventual georgic decline was marked by the emergence of analytic conceptions of increase and social diffusion. The transition was marked by the generational shift in influence from Davy to Justus von Liebig. Thus the practical aspect of cultivation (Whig agriculturalism) and the institutional initiatives associated with it move into the foreground. But until that happened, the idea of cultivation was more intimately connected with the conceptualisation of improvement and progress than with actual 'production'. Country argument was not only about metropolitan corruption and patriotic virtue, but also about the improvement of fertility. The 'Georgics of the Mind' was a Baconian formula that was picked up and developed by Scottish academic Whiggery. Georgic themes also formed an important part of the outlook of the English Country Whigs, which included chemical and geological preoccupations. The assumption by historians of political thought that the great landed aristocracy and the squirearchy were intellectually limited in comparison to the Scottish academy and the metropolitan salons needs to be amended. Historians have been looking in too few places. Georgic political economy was not the same as classical political economy but it reflected the scale of interests and priorities of a vigorous statesmanlike knowledge allied with science. Eventually, natural science came to be conscripted into analytic and compartmentalised forms, which had the effect of destabilising the older georgic tradition.

When the *Edinburgh Review* lauded the Lansdowne–Canning coalition of 1827 as 'the cause of liberty, the cause of toleration, the cause of political science', it testified to the rhetorical link between science and reform that was discernible by the time that the term 'liberal' entered public political discourse.[1] Isaac Kramnick has applied the term 'scientific liberalism' to the politics of Joseph Priestley, seeing a relentless and slightly sinister technocratic optimism in Priestleyan mechanics.[2] Liberalism might have had, as Jonathan Clark and others have argued, an essentially ecclesiastical provenance, but its connection with 'comprehension' was double-edged. Munificence and generosity of mind complemented an instinctively inclusive and latitudinarian sensibility.[3] From this perspective the adjectival antithesis of 'liberal' was not 'tory', 'conservative' or even 'arbitrary', but rather 'ignorant' (as an expression of narrowness). All Whig statesmen, including Foxites and Holland House paternalists who were suspicious of liberal political economy, liked to think of themselves as liberal in this sense. Other commitments conceived as conventionally 'liberal' might or might not be superimposed. The concern of much intellectual history has been to demonstrate how Whiggism was distinct from *laissez-faire* liberalism.[4] This book gives further weight to the notion that another key question was how to display one's comprehension. The political and moral sciences offered one rich set of resources. Another field was opened up by the natural sciences. Scientific liberality was no architectonic, ideological artefact, like the later 'Scientific Socialism'. It was, instead, a form of show. The Liberal Tory inheritors of Pitt had the capacity to compete over this specific ground, to co-opt and transform it. In contrast, high tories such as Eldon possessed a very different stylistic politics, resting on the demonstration of orthodox piety and loyalty, rather than comprehension or inclusion. They fell outside the fold of liberality. Romantic conservatives might be engrossed in natural philosophy, but liberality was not their concern. The Coleridgean clerisy was not liberal – its knowledge was esoteric, dialectical or mysterious. Romantic radicalism could, however, show affinities with scientific liberality. Ralph O'Connor hints as much in his discussion of Byron's open use of Cuvier compared with the hidden incorporation of geological allusion in Wordsworth, Coleridge and Keats.[5]

Thus scientific liberality had only a tangential relationship with positions commonly considered classically liberal, free trade for example, or the minimal state. Munificence of mind mattered a great deal more in the depiction of Whig statesmanship, penetrating and transforming political traditions inherited from the repository of political thought, such as the georgic and the republican. From the point of view of liberality, temporal characteristics usually taken as defining Whiggism such as teleological history and progress complemented an attitude of munificence.

Georgic depiction has been explored in histories of art and agriculture, but less so in histories of politics, ideas and science. The republican style is much more familiar to intellectual historians following in the footsteps of J.G.A. Pocock, and its deployment in natural scientific argument is steadily becoming recognised.[6] However, historians of science have not properly registered the implications of republican polemic. A case study examining the republican scientific argument of Henry Brougham indicates the importance of such rhetoric in debates on induction. The deployment of republican rhetoric in scientific controversy by Scottish Whigs needs to be tied into the republican turn performed by the Foxite opposition from the 1790s onwards. It should be noticed that the manipulation of republicanism in the context of natural science served to modify its character. Cognition had not previously been the focus of republican argument, which rested originally upon the virtues connected with real property and arms.

Rational sociability was very important for Whig liberality. The object of the Cheshire Whig Club, for example, was to gather friends of liberty in a rational union that would enable men to exercise their independence publicly in order to further liberal principles.[7] According to the Earl of Albemarle, who presided at the Suffolk Fox dinner, the utility of such occasions was both social and intellectual. A union of minds elicited liberal political ideas, adding to politeness, so that men of different social statuses and levels of knowledge could come together to support the principles of the constitution.[8] Of course loyalist and tory associations were also significant, but asocial Whiggery was a contradiction in terms in a sense that detached loyalism (or indeed Benthamism) conceivably was not. Behind this lay the Rockinghamite conception of party, a connection of enlightened public men from which measures in the public interest would spring. The Whig sociable style drew on the Burkean idea of party, parliamentary 'Friendship', the county meeting and aristocratic patronage, especially through the medium of the salon. The important thing to note is that Whig participation in philosophic activity and idealisation of scientific sociability drew upon this mode of party sociability.

Theology was an inescapable axis of definition, even in the case of Whig statesmen like Lord Holland for whom religious demands were uncongenial. Natural religion was not the only position of theological significance on the Whig side, but it was the most ubiquitous reference and served to bring out temperamental and intellectual differences between statesmen. Henry Brougham maintained that natural theology was 'a legitimate branch of natural knowledge', its business being 'to deduce causes from effects till we come to the very First Cause'.[9] This book has sought to add to the extensive secondary literature on this topic by focusing on Whig debates around ontology and bodies.

So far as the natural sciences themselves are concerned, geology, physiology, medicine and chemistry (particularly in relation to agriculture) are more in evidence within this framework than physics and mathematics. The book has also had less to say about the social and political history of statistics, an area that has been comprehensively surveyed in other contexts.[10] Instances of scientific usage have been brought forward in order to illustrate the styles of liberality, sociability, cultivation and statesmanship. While patronage of statistics was part of the public stance of some Whig statesmen (for example, Lansdowne and

Milton/Fitzwilliam in the 1830s), it was less relevant to the values of sociability and generosity, and the older georgic tradition.

If there is a distinct farmyard air about some of the topics covered, then this reflects the agrarian interests of country Whigs. Rescuing the agrarianism of the Whig party is an important move in the revision offered here. Permeable boundaries existed between the physical, or natural, sciences and the moral. We find instances of medicine bracketed with political economy, for example, or chemical agriculture implicated in didactic genres of literature. While these phenomena are intrinsically interesting, it is their ascribed or actual role in the hands of politicians that ultimately concerns us. This perspective adjusts the history of science to the narratives of party and government. Two broad themes should be noted. These are, first, the precedence of sociability and manoeuvre over methodological purity (for example, in 'induction'); and second, the importance of the land in Whig stylisations of science. The dilettantism of politicians, and their vulgarisation and misappropriation of scientific ideas, were more prevalent than subtle inference and the careful working out of doctrines.

Location by style

It is worth spending a few pages to see how the argument fits into a broader theoretical discussion. The present Cambridge School orthodoxy in intellectual history can be described as post-Wittgensteinian insofar as it treats ideas according to their discursive uses.[11] Take, for example, L.S. Jacyna's work on Scottish academic and medical whiggism. For Jacyna, philosophical Whiggism is located in a discursive grid constituting a system of values and rhetorical justification.[12] Jacyna claims that his analysis is not intended to uphold any particular theoretical programme.[13] Despite this, Jacyna's work falls within the dominant framework of contemporary intellectual history. More particularly, Jacyna's approach is similar to the moderate linguistic turn of intellectual historians such as Quentin Skinner and John Pocock, in which the strategic intentions of actors can be disclosed by their discursive interventions. The popularity of this approach among historians derives not only from its fertility of method. The revolution wrought by the new intellectual history in applying instrumental methods to texts has had the invaluable effect of establishing criteria for deciding which items of evidence are most worthy of scrutiny.[14] Priority, for example, is suggested by repetition of argument or illustration in different contexts. The recycling of an item indicates its relative significance because it substantiates prevailing conventions. A second strategy of evidence concerns the form of document in which allusion appears. In particular, if references are made or recommended in didactic texts (lectures, reading lists, encyclopaedias) then they have a particular instrumentality.[15] Consequently, the study of second-order or derivative texts assumes a status hitherto neglected in the study of seminal authors.

The self-portrayal of this new intellectual history has concentrated on its emancipation from historical materialism, and also on its escape from the timeless procession of great thinkers.[16] Among other contexts, the approach makes sense as a response to the Namier–Butterfield debates over party and ideology in eighteenth-century Britain,[17] the stimulus here being the perceived inadequacy of the Butterfieldian response to Namierite structural analysis. According to Skinner, Butterfield sought merely to reiterate the importance of principled as opposed to self-interested motivations.[18] By this view, Butterfield restated a certain type of conscious volition over that of Namier's subjects, whose attitudes were only seemingly self-conscious as they ran along beaten paths.[19] Skinner has exposed the vulnerability of the Butterfieldian reiteration of principled motivation. He points out that such arguments will always fall to Namierite intimations of hypocrisy, the expectation of bad faith and the cant of party discourse.[20] Skinner's aim is to offer a rival and more powerful refutation. Namierism rests on a logical error because the range of actions open to political agents is always limited by the scope of available principles yielding plausible positive and negative descriptions for actions.[21] Ideas re-enter historical study as performative mechanisms, and it is the task of the critic to identify the nature of such intentional manipulations. With this in mind, Skinner argues that it is necessary for a political historian to be a historian of political ideas.[22]

Thus the ghost of Sir Lewis Namier, as well as that of Marx, hovers over the Cambridge School. One fundamental assumption, however, is shared by Namier, Butterfield and Skinner, namely that there is a categorical distinction between the distribution of status and resources by patronage and connection on the one hand, and political definition by principle (whether as justification or motivation) on the other. Yet this is plainly not the case. Patronage and connection were in themselves expressive social and ideological acts. Moreover, the meaning of patronage changed according to its circumstance and purpose and contemporaries struggled to interpret its significance at various points. The present author has previously explored this phenomenon in the collapse of the Lansdowne–Canningite coalition (1827–1828),[23] and another illustration can be drawn from a suggestive document, the memorandum drafted by Charles Babbage to Prime Minister Lord John Russell in December 1846.[24] Complaining of his lack of award or preferment, Babbage listed the appointments secured by his peers, ranging from a baronetcy for John Herschel to William Whewell's Mastership of Trinity College, Cambridge. In the final text, he imputed the bestowal of patronage on others to personal and political interest, implying that his own claims rested on his scientific achievements alone. Patronage here appears as impartial reward, attesting to the rationality and disinterest of government. The various applications put forward by Babbage were then enumerated, all of which had been turned down (with the partial exception of the Lucasian professorship). What is striking about the document in this context is the bitterness revealed by the initial draft compared to the final version. The draft suggests that in his approach Babbage considered two strategies, corresponding to different conceptions of what patronage signified. In the jettisoned sections, Babbage vented his frustration at the refusal of Whig administrations to reciprocate his own political efforts, principally in the Cambridge University contests of 1829 and 1830.[25] He considered protesting to Russell that those who honestly shared the Prime Minister's opinions were even less likely to receive consideration for their claims.[26] He implied that this was a shameful reversal, and that he had received no encouragement for his reforming efforts. The convention of expectation that Babbage thought about invoking was patronage as a signal of partisan solidarity. His choice of the impartial reward model indicated the prevalence of the transcendent function of natural science in political expression. Science should rise above politics, but at the same time it was made the ground of a claim to political patronage. Babbage had to choose between conventions of patronage to make his case. To do so, he had to make sense of the signals Russell was sending by the distribution of offices.

Speech has never been the only way that information has been carried across political structures, and operative conventions have been social as much as discursive practices. Pragmatic rules and etiquettes demarcated the boundaries that politicians experienced and were capable of adjusting with others. Such conventions are best thought of as components of style rather than as ideologies or coherent propositional sets. At the core of this book, therefore, lies a contention that Whig styles comprised the language and behaviour by which Whigs displayed themselves to the political world. This was not a purely intellectual phenomenon, although it obviously did overlap with ideology. Essentially, this identifies science in the economy of signals by which statesmen portrayed or were expected to portray their activity. The function of patronage in the distribution of information, as well as goods and status, should be acknowledged, among other forms of political communication.

Ernst Mayr has argued forcefully that it is mistaken to assume that a changing scientific sequence is the same in character as a changing political sequence.[27] Political pressures may be coincident with scientific changes, but they are not truly causative, and neither can explanation run from science to politics (because politics, according to Mayr, is marked by sharp discontinuities). Instead, what is notable is the contrast between political oscillation and the logic of scientific discovery. The latter is cumulative, building on each preceding step.[28] One might respond by pointing to the rich texture of analyses that incorporate political contexts, but this aesthetic observation sidesteps the argument. How, then, might science contribute to the study of party politics? Adrian Desmond seeks to integrate biology and politics by projecting the utility of scientific doctrines in securing commercial and political privileges.[29] The argument here is that there is another fruitful answer in the expression of public styles for the purposes of political selection.

The concept of style has been under-theorised in high political narrative, even though it appears to apply readily and plausibly to it.[30] By contrast, an appreciable theoretical analysis has developed on the theme in the history and philosophy of science. This has focused on the coherence of the research school as a unit of analysis, the transferability of a category drawn from high aesthetics to epistemological definition, and the identification of national styles in scientific enquiry.[31] Although the preoccupations of high politics are quite distinct, the lines of this separate disciplinary debate are helpful in indicating the advantages and limitations of the concept.

Lorraine Daston and Michael Otte depict the concept of style as both elusive and indispensable.[32] Their own survey suggests that this description is both too ethereal and too vaunting. Scientific style is usually understood, they write, as a pattern arising during scientific work characteristic of an individual, school or nation.[33] The meaning is fairly clear, although the principal difficulty arises in identifying the unit of analysis, which can very easily become an exercise in circular reasoning. This weakness is most apparent in the treatment of national style, where the empirical connections between agents are the most indirect and diffuse.[34] The thicker the density of inter-reference and communication within a stratum, the more defensible it is to identify a style characteristic of it; hence the relative success of the analysis of research schools. Yet to argue that stylistic treatments are indispensable is too ambitious. Style is one element among others, a means of organising the remains of cultural communication. There is no evident *a priori* reason why a stylistic analysis should be preferred to, say, a class analysis; the success of any such exercise needs to be judged by the extent to which relationships are revealed by new juxtapositions. What style enables us to capture, as Gerald Geison notes, is the aspect of identification and display. From the stance of a study attempting to bring together high politics, science and the history of ideas, this is a promising perspective, so long as its limitations are understood.[35]

Clearly, the frameworks or guiding problems of political history differ markedly from those of the sociology of science debates.[36] The latter have focused on the laboratory construction of experimental authority, the process of theory prevalence (for example, through social networks) and the rhetoric of discovery addressed to specialist and popular audiences. By contrast, in order to integrate science and high political history, it is necessary to show how scientific activities and doctrines were capable of expressing power in spheres of legislation, party formation and social obligation. It is generally accepted that society and politics are contexts of science, though some contemporary scientists imply that these phenomena merely supply period detail to the self-referential emergence of scientific knowledge.[37] Here it is posited that science has partially expressed political culture through style.

The expressing or framing of politics by science specifically evokes the absorption of scientific engagement into the rules, or conventions, by which political persons acted. This prompts the thought that political style is a process rather than an object. A network of politicians manipulating science is plainly not the same as a research school because political stylisation is directed to other ends. The research school is entirely constituted by its institutional realisation and practice, while the political group looks to the exercise of power in the public arena. Rather than adding another scheme of classifying politicians, stylisation describes in the first place a process of putting science to work. Thus we need to go on to examine the relevant ends that various Whig statesmen set themselves in different degrees at different times: the political society that they wished to embody, the polemical battles that they wanted to win, the type of administration that they wanted to lead, the land they desired to represent and the religion that they wished to evoke. How could they form effective groups to attain these ends?

Historians of science have recently paid increasing attention to the relationships of friendship, interest and rivalry among the users and consumers of natural philosophy. A striking example has been furnished by James Secord.[38] In his account of the 'Aristocracy of Nature' who had to deal with the *Vestiges of the Natural History of Creation* (1844), Secord identifies a varied pattern of responses among different Whig and Peelite networks, connecting active experimenters such as Viscount Morpeth and the Earl of Rosse, institutional forums such as the Royal Society and British Association, and circles of discussion at Holland House, Lady Palmerston's Cambridge House and the Drayton manor of Sir Robert Peel. This kind of study introduces topics that are familiar to political history – patronage and connection, 'ins' and 'outs' – especially if interactions are conceived as flows of information, as well as of power, status and assets. Signals of favour and disfavour, encouragement and restraint, are produced by the manipulation of resources, and one does not need the behaviourist psychology employed by Sir Lewis Namier to acknowledge his main insight, the centrality of acquisition, gifts and exchanges in the dissemination of influence.[39]

The informative character of patronage indicates that it is misplaced to think of political groups as passive frameworks, within which discourses operate.[40] Historians of science such as Desmond have sought to incorporate social constraints into their accounts of the views of nature held by past societies.[41] This should prompt historians of political ideas to confront their own ghostly and disembodied histories, for example by relating textual genealogies in the history of political thought to social and high political imperatives. Meanwhile, a different sort of internalism has characterised those political histories that emphasise the closed conventions of the Westminster game.[42] Styles not only locate political actors discursively, but also signal their positions in social networks. To pursue this approach entails a definite shift in focus towards the formations that mapped the environment of early-nineteenth-century whiggery.

Politics as artificial selection

The argument in the *Origin of Species* (1859) begins with a discussion of artificial selection.[43] Understandably, biological and social theorists have treated this as a curtain-raiser to the main event, an analogical introduction to the theory of natural selection.[44] Yet the originality and profundity of Darwin's central insight has overshadowed the usefulness of his initial move. The notion of selection puts the study of Whig manners into an interesting theoretical context. Rather than trying to stretch natural selection to cover political phenomena it is more fruitful to compare the framework of human choice that Darwin did discuss.

In one of its aspects, politics can be understood as an artificial selection of groups or associations: as a process of political selection.[45] Of course the sentient and reciprocal means by which Whigs attempted to cultivate other Whigs was vastly more complicated than their dog- or sheep-breeding. We have to be careful to qualify what we are saying here. Many men across the political spectrum had an interest in animal breeding or plant cultivation: for example, of the gentleman cultivator politicians cited by Darwin, Lord John Charles Spencer (Viscount Althorp) and Sir John Sebright could fairly be described as Whigs, while Lord Somerville had enjoyed the patronage of George III (himself a prodigious breeder) and of Sir Joseph Banks.[46] Whigs seldom spoke about their political relationships in terms of domestication – to do so would have been uncouth and insulting (but recall Sydney Smith's quip, discussed by Walter Bagehot, that Lord Lansdowne 'looks for talents and qualities amongst all ranks of men, and adds them to his stock of society, as a botanist does his plants').[47] In the *Origin* Darwin does not touch upon political manoeuvre or parliamentary manipulation, and there is little evidence at this stage that he intended his argument to extend there. Nevertheless, the central observations Darwin made of artificial selection do throw light on some of the things I have been describing in this book.

Darwin argued that in selecting, men act on external and visible characters.[48] This was as at least as true of political recognition in the nineteenth century as of the identification of desirable traits in livestock. The styles and signs we have been discussing helped Whigs to decide whom to promote and aid. Like all political agents, Whigs only had access to interior motivations through visible literary productions, connections or political actions. These signs were useful in political selection. Unlike the artificial selection of flora and fauna, political selectees were agents too, and so were able to respond rapidly to the observed economy of signals and to select in turn. Political evolution (political descent with modification according to changed circumstances of selection) proceeded more rapidly than in the turnover of natural generations, although generational change (for example, from Fox's original cohort to the Young Whigs) could also be observed.

We need to deal with an objection, which might be put as follows. Artificial selection, in the sense that Darwin discussed it, has a genuinely strong analogy with natural selection because both the human domestication of nature and natural selection share a common and constant substrate, namely plant and animal organisms. Twentieth-century science has defined the common substrate even more rigorously in terms of the DNA environment and genetic expression. Artificial selection of flora and fauna (or nowadays their genes) looks as if it bears a strong analogy with political selection, because both deal with structures of human choice, but in fact the analogy is weak because, first, they do not share a common substrate and, secondly, because politics itself has a range of substrates – policies, parties, electoral contests and so forth. The same difficulty as with 'meme' theory rears its head. There is really no single-shared plane of human action but a profusion of different cultural channels with different properties, and the boundaries of the evolving entities within these channels are opaque.[49] How then do we explain as a constant object the substrate of politics, and as a constant subject the units of cultural evolution?

The answer is that we cannot. The objection is fatal to the 'meme' theory, while it properly limits the theory of artificial political selection. The meme theorist has not only to define the essential similarity between, for example, a tune and a political slogan, but also to assert the explanatory significance of this unity. However, the historian of political selection can say that politics is composed of temporary and overlapping substrates and subjects, and that his or her field consists of a series of evolutionary runs. The problem has a bearing both on the choice of topics of study (what makes a particular political substrate?) and on the question of complexity. An approach based on political selection will be useful for a certain set of historical problems and trivial for others, depending on how clearly a temporary common substrate or subject can be identified. Questions of the propagation of political groups and identities (for example, Whig manners) fruitfully accord with the study of political selection; other problems may not. So far as complexity is concerned, evolution leads us to expect that late forms which have undergone extended modifications will be more complex than earlier ones. As Pocock illustrates, there is a tangible way in which we can say that the Whiggery of the Reform Act period was more complex than the Whiggery of the seventeenth-century Exclusionists. There are more ways of being a Whig in the later period and Exclusionism is folded into the history of Reform Whiggery.[50] Yet most political historians would baulk at declaring that the politics of the Liberal fragments of the 1930s was more 'complex' than nineteenth-century Liberalism.[51] Political extinctions and contractions do occur, and one way of thinking about this is as a process of substrate or subject abandonment. The clock of political evolution is started again and again, in an overall pattern of recurring complexity.

The most important distinction between artificial and natural selection is not the question of substrates or discrete subject boundaries, but the distinction of *scarcity*. For this, we have to turn to Darwin's inspirer Thomas Malthus. Environments that provide scarce resources relative to populations will produce the kind of competitive selection of variations seen under natural selection. In politics there certainly are moments of Malthusian scarcity – wars and revolutions – where the political environment will trigger acute struggles for survival and competitive pressure analogous to natural selection. But for most of the time political selection (particularly as practised among elites) occurs in situations of relative *abundance*. Through their patronage and organisation, statesmen are able to create abundant micro-climates. The more elitist the structure of politics (that is, the more statesmen are able to command large resources and protect their clients) the less Malthusian the pattern of politics becomes. Competition certainly exists, but it is stylised, sometimes ritualised, *agreed* competition, of the type obtaining between rival pigeon-fanciers or sheep-exhibitors. Most of the time, the Whig stylistic processes we have looked at were closer to this sort of domesticating rivalry than to the Malthusian struggle for subsistence.

Darwin saw how artificial selection was often carried out in conditions of imperfect knowledge and power, and this is an invaluable corrective for intellectual historians.[52] Certainly, there were outstandingly self-aware and reflective Whig intellectuals. Yet Whigs would sometimes use ideas or respond to concrete political problems without grasping all their implications or really knowing what they were talking about. Here bring to mind Goldman's observation that Whig politicians in the 1830s patronised statistical study without being fully conversant with all the methodological debates.[53] Whig politics did not evolve according to a grand plan (certainly not to an overarching teleology) even though various Whig intellectuals anxiously scrutinised their history to find such a trajectory. Instead, a series of challenges, year after year, modified Whig political practice. Whigs would emulate one another's practices to meet concrete needs: for example, sociable gatherings were organised at Holland House and at Bowood to gather ideas and project influence. To use the Darwinian lexicon, artificial selection could be methodical or unconscious.[54] The effect over a long run was a course of accumulated adaptations.

The framework of artificial selection is compatible with any number of motivations and the study of ideology and political ideas remains indispensable to make sense of political selection. To speak of purposes in natural selection is misleading and mistaken but the opposite is true for artificial selection. Sir Lewis Namier, who effectively described how some means of political selection operated, was still wrong to minimise the role of doctrine. The current lack of agreement about the way in which discourse and language work perhaps puts us in a position similar to that of Darwinists before the Modern Synthesis. Should theorists of discourse ever come to a consensus about their field, we would be able to extend the theory of political selection, but we do not have to wait for such an elusive breakthrough in order to discuss concrete examples of political domestication. Discursive principles might turn out to be quite compatible with a restricted view of the role of individual agents and the unfinished, open and provisional character of discourses. As Darwin argued, both breeds and dialects lack a definite point of origin.[55] Darwin posited a continuum between varieties and species, refusing rigid and final definitions. The continuity as well as the variety of political positions accords well with this approach.

Darwin noted how large variations would be produced by the artificial selection of slight differences. The striking range of human political societies, styles and ideologies has rendered detailed cultural explanations based on natural selection implausible. In the case of the Whig styles examined here, we can ask, why so many whiggeries, and not the expansion of one to fill a political niche? The problem is lessened if we take Darwin's argument that domestication tends to mould the whole of a given subject more sporadically and unevenly than natural selection does.[56] This is another way of saying that artificial selection in conditions of relative abundance makes better sense of the breadth of surplus political forms. In conclusion, it is useful for political historians to work within a Darwinian evolutionary framework so long as they do not fall into the trap of thinking that they are dealing in an immediate way with natural selection.[57] Artificial selection is the more relevant of Darwin's ideas.

Implications

The recent historiography of early-nineteenth-century liberalism has been analytic in its aspiration to distinguish more clearly the Whig conversation of interests from political individualism,[58] to discern between Whig paternalism and social and economic liberalism,[59] and to separate liberal Anglicanism from evangelical and High Church identities.[60] By contrast, the present argument has pursued a synthetic agenda, seeing in scientific liberality and other manners a form of common criteria for public display in order to facilitate political selection. Engagements with natural philosophy displayed breadth of knowledge which was generous and rational. This facilitated Whig intellectual sociability and coalition building. By the middle of the nineteenth century, however, this sort of liberality had had its day. New forms of sensibility arose; cultivation was outgrown by concepts of serial industrial replication while political comprehension was outpaced by moral dialectics.

This book is based in the first place, therefore, on a refusal to admit with Richard Brent that the term 'liberal' can be used interchangeably with 'progressive'.[61] It has also stepped back from the common assumption that liberalism's premises lay in the *laissez-faire* advocacy of individualism and limited public intervention (whether Benthamite or evangelical). Instead, the argument has taken as its point of departure the Clarkian idea of liberalism as originating in comprehension, generosity and liberality, and has linked it to science through ostentatious munificence.[62] While Clark's intention was to show the weakness of liberal conceptions (as compared, for example, to radicalism), this book has demonstrated their presence across a series of Whig depictions. This mode was stylistic, rather than directly ideological, suggesting that we ought to consider the Duke of Bedford's sheep-shearing in the same framework as the literary georgic arguments of Dugald Stewart. The argument has had more to say about manners and sociabilities than about the world of experiment – except agrarian experiment – and it has tended to scrutinise reading lists and patronage connections rather than laboratories. When the focus of knowledge moved from the estate into the laboratory and the research school, this was not simply one stage in a narrative of professionalisation or institutional specialisation; it also compromised scientific liberality and the georgic political portrayal. Meanwhile, the iconic manipulation and co-option of mascot chemists (Davy and Liebig) was a typical move by Whig statesmen.

The display of scientific liberality allowed Whigs to draw on their style of political sociability and the pursuit of voluntary assent, both for vertical leadership roles (for instance, the county meeting and the hierarchical club for Broughamite educational institutions) and for horizontal alliances (the model of party and coalition transposed to scientific societies formed with peers). The coalition of knowledge, with its basis in the sociability of party, was one element in Whig style. It differed from the template of enlightened philosophical society, not least in its greater flexibility and toleration towards diversity of opinion held together by shared core principles. That is, its aim was not to forge unanimity but to manage difference, trusting more to the process of common comportment than to the consensus that might emerge through it. Institutions such as the London Geological Society witnessed this sort of sociability. The significance of Baconianism lay in the collaborative character of its ideal programme, which dovetailed with Whig conceptions of the party and coalition of knowledge.

Henry Brougham's scientific republicanism was a peculiar form of rapprochement with the Foxite commonwealth register, whereas Horner's painstaking political science, and pursuit of the currency question, placed him alongside magisterial liberals such as Lansdowne and Grenville. The magisterial deployment of scientific liberality filled the technical vacuum left by the Rockinghamite–Foxite tradition. Such manoeuvres are best seen as selective acts of political display rather than as scrupulous methodology. Within Whiggery, progressive history and the guardianship of improvement formed a common motif: but the possibility of decline, rather than the inevitability of ascent, was also an important element. It was possible for Whig statesmen to portray their activity within a series of timeless, ideal landscapes that were pastoral and harmonious in character. These georgic milieus were static in structure while being dynamic in content – labour and cultivation were permanent elements. This is not to say that the construction of these depictions was ahistorical or unchanging, or that the ideal landscapes of the 1850s were identical to those of the 1790s. Nevertheless, there is enough of a linkage and descent for us to be able to describe a long-lived phenomenon. The pragmatic chemistry of English Foxite aristocrats such as the Bedfords fell partially within this georgic matrix.

The scientific liberal style of rational speculation was compatible with a range of positions regarding body, mind and instinct. Brougham's exchanges with Althorp on dualism and instinct displayed a kind of contradictory escapism (escape from politics that was politicised through the evocation of disinterest). This paralleled the polite conversation and geological roaming of the Whig salons. While Brougham exhibited an intense form of devotion in a fundamentally heterodox natural theology, Althorp's orthodox religion allowed a theoretical relaxation towards materialism – a relaxation that was sustainable only because contingency in the universe was scripturally determined and metaphysics was therefore unlikely to prove subversive. However, scientific liberality was not the same as religious liberalism, as its aversion to mystery, esotericism and paradox demonstrates. Despite its basis in toleration and latitudinarian sensibilities, it was vulnerable to anti-Catholicism. The metaphysical challenges of Catholicism flanked the threat to Whiggery from radical and materialist doctrines. This could have political implications, and the episode of the Ecclesiastical Titles controversy was an indication of the limits of liberality's liberalism.

Notes

1 Manners, science and politics

1. J. Bord, 'Whiggery, science and administration: Grenville and Lord Henry Petty in the Ministry of All the Talents, 1806–7', *Historical Research*, 76 (2003), 108–127.
2. L.G. Mitchell, *The Whig World, 1760–1837* (London: Hambledon and London, 2005); *Holland House* (London: Duckworth, 1980).
3. Henry Edward, Lord Holland (ed.), Henry Richard Vassall, Lord Holland, *Memoirs of the Whig Party During My Time* (London: Longman, Green, Brown and Longmans, 1852), I, pp. 45–55; Lord Stavordale (ed.), Henry Richard Vassall, Lord Holland, *Further Memoirs of the Whig Party with Some Miscellaneous Reminiscences, 1807–1821* (London: John Murray, 1905), pp. 370–375.
4. H. Brougham, *Discourse of the Objects, Advantages and Pleasures of Science* (2nd edn, London: Baldwin and Cradock, 1827), p. 6.
5. R. Yeo, *Defining Science: William Whewell, Natural Knowledge and Public Debate in Early Victorian Britain* (Cambridge: Cambridge University Press, 1993), pp. 28–48, see p. 29.
6. See J. Parry, *The Rise and Fall of Liberal Government in Victorian Britain* (New Haven and London: Yale University Press, 1993), pp. 73–75, for an overview; E.A. Wasson, *Whig Renaissance: Lord Althorp and the Whig Party, 1782–1845* (New York and London: Garland, 1987); 'The coalitions of 1827 and the crisis of Whig leadership', *Historical Journal*, 20 (1977), 587–606.
7. Parry, *Rise and Fall of Liberal Government*, p. 167; L.G. Mitchell, *Charles James Fox* (London: Penguin, 1997), p. 195. Fifty-two self-described 'Whigs' were returned in 1847, while Fox identified sixty-nine Foxites in 1802.
8. See L.G. Mitchell, 'Foxite politics and the great reform bill', *English Historical Review*, 108 (1993), 338–364; *Lord Melbourne, 1779–1848* (Oxford: Oxford University Press, 1997), pp. 3–40, 142–210.
9. C. Geertz, *The Interpretation of Cultures: Selected Essays* (London: Fontana, 1993), pp. 12–13.
10. N. Gash, *Reaction and Reconstruction in English Politics, 1832–1852* (Oxford: Clarendon Press, 1965), esp. chs. 5–6, pp. 119–200; *Aristocracy and People, Britain 1815–1865* (London: Edward Arnold, revs. edn 1983), pp. 156–186.
11. P. Mandler, *Aristocratic Government in the Age of Reform: Whigs and Liberals, 1830–1852* (Oxford: Clarendon, 1990).
12. R. Brent, *Liberal Anglican Politics: Whiggery, Religion and Reform 1830–1841* (Oxford: Clarendon, 1987).
13. D. Forbes, *The Liberal Anglican Idea of History* (Cambridge: Cambridge University Press, 1952).

14. J.W. Burrow, *Whigs and Liberals: Continuity and Change in English Political Thought* (Oxford: Clarendon, 1988).
15. Signally in J.G.A. Pocock 'The varieties of Whiggism from Exclusion to Reform: a history of ideology and discourse', *Virtue, Commerce, and History* (Cambridge: Cambridge University Press, 1985), pp. 215–310.
16. Ibid. and J.G.A. Pocock, *The Machiavellian Moment: Florentine Political Thought and the Atlantic Republican Tradition* (2nd edn, Princeton, NJ and London: Princeton University Press, 2003), pp. 546–547.
17. B. Hilton, *A Mad, Bad, and Dangerous People? England 1783–1846* (Oxford: Clarendon Press, 2006), pp. 309–371, 439–492, esp. pp. 439–441, in the context of 'The politics of anatomy and an anatomy of politics, c.1825–1850', in S. Collini, R. Whatmore, and B. Young (eds), *History, Religion and Culture: British Intellectual History, 1750–1950* (Cambridge: Cambridge University Press, 2000), pp. 179–197; *Corn, Cash, Commerce: The Economic Policies of the Tory Governments, 1815–1830* (Oxford: Oxford University Press, 1977), pp. 303–314; and especially *The Age of Atonement: The Influence of Evangelicalism on Social and Economic Thought, 1785–1865* (Oxford: Clarendon Press, 1991), pp. 147–162.
18. T.A. Jenkins, *Gladstone, Whiggery and the Liberal Party, 1874–1886* (Oxford: Clarendon, 1988).
19. Parry, *Rise and Fall of Liberal Government*, pp. 167–303, esp. pp. 274–289.
20. Ibid., p. 1n.
21. Gash, *Reaction and Reconstruction*, pp. 157–200, esp. pp. 157–169. Leslie Mitchell points to a pattern set before the French Revolution: *Charles James Fox*, 'In Foxite Society', pp. 94–107.
22. T. Macaulay, 'War of the succession in Spain', *Edinburgh Review* [January, 1833], in A.J. Grieve (ed.) *Critical and Historical Essays by Thomas Babington Macaulay* (2 vols, London: Dent, 1907), II, p. 111.
23. Hilton, 'The politics of anatomy and an anatomy of politics', pp. 190–194.
24. W.H.G. Armytage, 'Charles Watson-Wentworth, Second Marquess of Rockingham, F.R.S. (1730–1782): some aspects of his scientific interests', *Notes and Records of the Royal Society of London*, 12 (1) (1956), 64–76.
25. I. Kramnick, *Bolingbroke and his Circle: The Politics of Nostalgia in the Age of Walpole* (Cambridge, MA: Harvard University Press, 1968), p. 139. For Bolingbroke's rejection of empiricism, see ibid., p. 45. Kramnick's point is consistent with Reed Browning, *Political and Constitutional Ideas of the Court Whigs* (Baton Rouge, LA and London: Louisiana State University Press, 1982); Caroline Robbins, *The Eighteenth-Century Commonwealthsman* (Cambridge, MA: Harvard University Press, 1959); H.T. Dickinson, *Walpole and the Whig Supremacy* (London: English Universities Press, 1973) and *Liberty and Property: Political Ideology in Eighteenth Century Britain* (London: Weidenfeld and Nicolson, 1977).
26. Reed Browning indicates the Ciceronian structure of Court Whig ideology, *Political and Constitutional Ideas of the Court Whigs*, pp. 175–256, particularly of the court conception of the balanced constitution, ibid., pp. 245, 252; J.H. Plumb, *Sir Robert Walpole: The King's Minister* (London: The Cresset Press, 1960); J.B. Owen, *The Rise of the Pelhams* (London: Methuen and Company, 1957); R. Browning, *The Duke of Newcastle* (New Haven CT,

and London: Yale University Press, 1975); S. Ayling, *The Elder Pitt, Earl of Chatham* (London: Collins, 1976).

27. L.S. Jacyna, *Philosophic Whigs: Medicine, Science and Citizenship in Edinburgh, 1789–1848* (London, New York: Routledge, 1994), p. 31.
28. Dickinson, *Liberty and Property*, p. 145.
29. Franklin to Cadwallader Evans, 7 September 1769, mentioned in V.W. Crane, 'The Club of Honest Whigs: *The William and Mary Quarterly*, 3rd. Series 23 (1966), 210–233, 211.
30. Ibid.; and see for context Robbins, *Eighteenth-Century Commonwealthsman*, pp. 320–377.
31. D.P. Miller, 'The "Hardwicke Circle": the Whig supremacy and its demise in the 18th-century Royal Society', *Notes and Records of the Royal Society of London*, 52 (1998), 73–91.
32. Ibid., 81–83; S. Schaffer, 'The consuming flame: electrical showmen and Tory mystics in the world of goods', in R. Porter and J. Brewer (eds), *Consumption and the World of Goods* (London: Routledge, 1993), pp. 489–526.
33. On the Whig side, see M. Jacob, *The Newtonians and the English Revolution, 1689–1720* (Hassocks: Harvester Press, 1976), and for an example of Newtonian Hanoverianism, see J.T. Desaguliers, *The Newtonian System of the World, the Best Model of Government: An Allegorical Poem* (Westminster, 1728), cited in Paul Langford, *A Polite and Commercial People: England, 1727–1783* (New York and Oxford: Oxford University Press, 1992), p. 11. For Toryism, see A. Guerrini, 'The Tory Newtonians: Gregory, Pitcairne and their circle', *Journal of British Studies*, 25 (1986), 288–311.
34. For this parallel between John Trenchard and Walter Moyle (d.1721) and Bolingbroke, see Isaac Kramnick, *Bolingbroke and his Circle*, pp. 138–139, 255–256.
35. M.C. Jacob and L. Stewart, *Practical Matter: Newton's Science in the Service of Industry and Empire, 1687–1851* (Cambridge MA, and London; Harvard University Press, 2004), p. 74.
36. Ibid., pp. 18–25.
37. L. Stewart, *The Rise of Public Science: Rhetoric, Technology and Natural Philosophy in Newtonian Britain, 1660–1750* (Cambridge: Cambridge University Press 1992), pp. 90–92, 109–111, 118, 189–190, 206.
38. Ibid., pp. 80–85, 203, 335.
39. Ibid., pp. 266–272, 317–326.
40. Ibid., p. 16.
41. Ibid., pp. 156, 392.
42. Ibid., pp. 153–154.
43. P.N. Miller (ed.), *Joseph Priestley: Political Writings* (Cambridge: Cambridge University Press, 1993), pp. xv–xvi.
44. John Norris, *Shelburne and Reform* (London: Macmillan and Company, 1963), pp. 82–99. Shelburne is referred to here occasionally as such even after his elevation to the marquessate in 1784, where this is necessary to distinguish him from Henry Petty-Fitzmaurice, the third Marquess of Lansdowne (1780–1863), who in turn was known as Lord Henry Petty until his succession in 1809.
45. Robert Stewart, *Henry Brougham, 1778–1868: His Public Career* (London: Bodley Head, 1985), p. 8.

46. Leonard Horner (ed.), *Memoirs and Correspondence of Francis Horner MP* (2 vols, 2nd edn, London: John Murray, 1853), I, pp. 209–210.
47. Ibid.
48. Gash, *Reaction and Reconstruction*, p. 160.
49. Mitchell, 'Foxite politics and the Great Reform Bill', 338–364.
50. See Wasson, 'Coalitions of 1827', and J. Bord, 'Our friends in the north: patronage, the Lansdowne Whigs, and the problem of the liberal centre, 1827–8', *English Historical Review*, 117 No. 470 (2002), 78–93.
51. Wasson, *Whig Renaissance*; 'Coalitions of 1827'.
52. Austin Mitchell, *The Whigs in Opposition, 1815–1830* (Oxford: Clarendon Press, 1967), mounts a standard survey, principally in Chapters I–III, pp. 1–81.
53. For an overview, see Hilton, *A Mad, Bad and Dangerous People?* pp. 348–350.
54. Mitchell, *Holland House*, passim.
55. H. Grey Bennet, 'Diary of the House of Commons', Henry Grey Bennet Papers, London, House of Lords Record Office (HLRO), HL/PO/RO/1/129.
56. D. Rapp, 'The left-wing Whigs: Whitbread, the mountain and reform, 1809–1815', *Journal of British Studies*, 21(2) (1982), 35–66.
57. Grey Bennet, 'Diary of the House of Commons', HL/PO/RO/1/129, pp. 31–32, 41–42.
58. Hilton, *Corn, Cash, Commerce*, pp. 13–14.
59. Bennet, 'Diary of the House of Commons', HL/PO/RO/1/129, esp. pp. 85, 100–101.
60. L.G. Mitchell, *C.J. Fox and the Disintegration of the Whig Party* (London, Oxford University Press, 1971), p. 238.
61. J.R. Dinwiddy, *Radicalism and Reform in Britain, 1780–1850* (London: Hambledon, 1992), p. 13.
62. Mitchell, *Whigs in Opposition*, p. 17.
63. Ibid., p. 3.
64. Ibid., p. 13.
65. Ibid., pp. 9–10.
66. Burrow, *Whigs and Liberals*, p. 2.
67. Brent, *Liberal Anglican Politics*, especially 'The political and intellectual origins of Liberal Anglicanism: from Foxism to constitutional Moralism', pp. 19–64.
68. Pocock, *Virtue, Commerce and History*, pp. 218, 279–310, esp. p. 285.
69. Biancamaria Fontana, *Rethinking the Politics of Commercial Society: The Edinburgh Review, 1802–1832* (Cambridge: Cambridge University Press, 1985), pp. 6, 72–76, 105–110, 140–146.
70. The ethos of anti-court aristocratic politics is comprehensively described by Leslie Mitchell, *C.J. Fox and the Disintegration of the Whig Party*, p. 238; see also his *Holland House* and 'Foxite politics and the great reform bill', mentioned previously.
71. Mitchell, *Holland House*, p. 184. See particularly Lord Holland's insistence to Lord Greville that political progress could be halted, with catastrophic results: F.M. Bladon (ed.) *The Diaries of Robert Fulke Greville*, vol. III (London: Bodley Head, 1930), 9 January 1835, pp. 197–198 discussed in ibid.
72. Brent, *Liberal Anglican Politics*, p. 43.

73. Mandler, *Aristocratic Government*. The political taxonomies of Brent and Mandler are compared by Boyd Hilton, 'Whiggery, religion and social reform: the case of Lord Morpeth', *Historical Journal*, 37 (1994), 829–859, esp. 829–839.
74. Hilton, *The Age of Atonement*, pp. 237–248; Wasson, *Whig Renaissance*, pp. 126–127.
75. J.C.D. Clark, 'A general theory of party, opposition and government, 1688–1832', *Historical Journal*, 23 (1980), 307–308.
76. *Edinburgh Review or Critical Journal* [London: Longmans, Green and Company, 1802–1929], XXXIX (1812), 2, 35, discussed in John Clive, *Scotch Reviewers: The Edinburgh Review, 1802–1815* (London: Faber and Faber, 1957), p. 119; Mitchell, *Whigs in Opposition*, p. 24.
77. See, for example, Smith to Lady Grey, 7 February 1835, in N.C. Smith, *Selected Letters of Sydney Smith* (London, Oxford University Press, 1956), p. 232; Lord Russell to Lord Melbourne, 9 September 1839, mentioned in Gash, *Reaction and Reconstruction*, p. 157.
78. Memorandum of Lord Clarendon discussed by Gash, *Reaction and Reconstruction*, p. 192.
79. Russell to Tierney, 26 November 1798, Winchester, Hampshire Record Office 31 M70/Item 60.
80. Horner to Jeffrey, 15 September 1806, in K. Bourne and W.B. Taylor (eds) *The Horner Papers: Selections from the Letters and Miscellaneous Writings of Francis Horner, MP, 1795–1817* (Edinburgh: Edinburgh University Press, 1994), p. 427.
81. Ibid., p. 428.
82. Smith to Lady Grey, 7 February 1835, Smith, *Selected Letters of Sydney Smith*, p. 232.
83. Russell to Melbourne, 9 September 1839, mentioned in Gash, *Reaction and Reconstruction*, p. 157.
84. Scarlett to Lord Milton, 20/21 April 1827; see E.A. Smith, *Whig Principles and Party Politics: Earl Fitzwilliam and the Whig Party, 1748–1833* (Manchester: Manchester University Press, 1975), p. 385.
85. H. Brougham, *Dialogues on Instinct with Analytic View of the Researches on Fossil Osteology* (London: C. Knight and Company 1844), p. 13. Studies of Brougham include A. Aspinall, *Lord Brougham and the Whig Party* (Manchester: University of Manchester, 1927); C.W. New, *The Life of Henry Brougham to 1830* (Oxford: Clarendon Press, 1961); Stewart, *Henry Brougham*; and T.H. Ford, *Chancellor Brougham and His World: A Biography* (Chichester: Barry Rose Law, 2001).
86. Brougham, *Dialogues on Instinct*, p. 16.
87. Most recently in an excellent article, J.F.M. Clark, 'History from the ground up: bugs, political economy and God in Kirby and Spence's *Introduction to Entomology*, 1825–1856', *Isis*, 97 (2006), 28–55, esp. 45–47. See R.J. Richards, *Darwin and the Emergence of Evolutionary Theories of Mind and Behaviour* (London and Chicago: University of Chicago Press, 1987), p. 139. For an interesting variant that takes Brougham's classicism seriously, see C.D. Pearce, 'Lord Brougham's neo-paganism', *Journal of the History of Ideas*, 55 (1994), 651–670.
88. Brougham, *Dialogues on Instinct*, p. 15.

89. Ibid.
90. See, for example, John G. McEvoy, 'Positivism, Whiggism, and the chemical revolution: a study in the Historiography of Chemistry, *History of Science*, 35 (1997), 3, 26–27.
91. Sir Herbert Butterfield, *The Whig Interpretation of History* (London: G. Bell and Sons, 1931). This text, and its treatment in M. Teich and R. Young (eds), *Changing Perspectives in the History of Science: Essays in Honour of Joseph Needham* (London: Heinemann Educational, 1973), pp. 127–147, is the basis of the extended definition of Whiggism in Chris Wilde's influential entry on 'Whig history', in W.F. Bynum, E.J. Browne, and R. Porter (eds), *Dictionary of the History of Science* (London and Basingstoke: Macmillan, 1981), pp. 445–446; praised by M.J.S. Hodge, 'The history of the earth, life and man: Whewell and Palaetiological science', in M. Fisch and S. Schaffer (eds), *William Whewell, A Composite Portrait* (Oxford: Clarendon, 1991), p. 255 and n. 1. The received definition is repeated by Peter J. Bowler in his review, 'The Whig interpretation of Geology, *Biology and Philosophy*, 3 (1988), 100. For attempts to rehabilitate 'Whig' narratives see A. Rupert Hall, 'On Whiggism', *History of Science*, 21 (1983), 45–59; Ernst Mayr, 'When is Historiography Whiggish?', *Journal of the History of Ideas*, 51 (1990), 301–309.
92. R. Yeo, *Defining Science*, pp. 168–169; J.W. Burrow, *A Liberal Descent. Victorian Historians and the English Past* (Cambridge: Cambridge University Press, 1981).
93. M. Norton Wise with Crosbie Smith, 'Work and waste: political economy and natural philosophy in nineteenth century Britain (I)', *History of Science*, 27 (1989), 263–301, 267. Parts II and III of this study are in the same journal, 391–449 and 28 (1990), 221–261.
94. S.F. Cannon, *Science in Culture: The Early Victorian Period* (Folkestone, Dawson, 1978), pp. 29–71; Hilton, *Age of Atonement*, p. 30; J.A. Secord, *Victorian Sensation: The Extraordinary Publication, Reception, and Secret Authorship of Vestiges of the Natural History of Creation* (Chicago: University of Chicago Press, 2000), p. 410.
95. P. Corsi, 'The heritage of Dugald Stewart: Oxford philosophy and the method of political economy', *Nuncius*, 2 (1987), 89–144, see 122–123.
96. Secord, *Victorian Sensation*, pp. 406–410.
97. S. Schaffer, 'The nebular hypothesis and the science of progress', in J.R. Moore (ed.), *History, Humanity and Evolution: Essays for John C. Greene* (Cambridge: Cambridge University Press, 1989), pp. 131–164.
98. A. Desmond, *The Politics of Evolution: Morphology, Medicine and Reform in Radical London* (Chicago and London: University of Chicago Press, 1989), p. 18.
99. Ibid., p. 93. Charles Bell helped Brougham to produce the middle volumes (1836) of an annotated edition of William Paley's *Natural Theology* [1802] between 1835 and 1839. The first, fourth and fifth volumes contained Brougham's own 'Discourse of natural theology', 'Dialogues on instinct' and 'Analytic view of the researches on fossil Osteology'. See Charles Bell and Henry Brougham (eds), *Paley's Natural Theology with Illustrative Notes* (5 vols, London: Charles Knight, 1836).

100. Desmond, *Politics of Evolution*, pp. 30–31.
101. W. Thomas, *The Philosophic Radicals: Nine Studies in Theory and Practice, 1817–1841* (Oxford: Clarendon Press, 1979), pp. 53–56.
102. Cannon, *Science in Culture*, pp. 251–252.
103. Roy M. Macleod, 'Whigs and savants: reflections on the reform movement in the Royal Society, 1830–48', in Ian Inkster and Jack Morrell (eds), *Metropolis and Province: Science in British Culture, 1780–1850* (London: Hutchinson, 1983), pp. 59, 68, 86 [n.57].
104. Ibid., p. 77.
105. Ibid., pp. 77–81, esp. p. 78. Cannon, *Science in Culture*, pp. 137–201, esp. pp. 138, 141, 169–170.
106. Ibid., pp. 57–61. See Simon Schaffer, 'The nebular hypothesis and the science of progress', in Moore (ed.) *History, Humanity and Evolution*, pp. 131–164.
107. L.S. Jacyna, 'Immanence or transcendence: theories of life and organisation in Britain, 1790–1835', *Isis*, 74(3) (1983), 329.
108. Outstanding works and collections on the period include Adrian Desmond, *The Politics of Evolution*; Jan Golinski, *Science as Public Culture: Chemistry and Enlightenment in Britain, 1760–1820* (Cambridge: Cambridge University Press, 1992); Secord, *Victorian Sensation*; Moore (ed.), *History, Humanity and Evolution*.
109. See, for example, Anne Hardy, 'Lyon Playfair and the idea of progress: science and medicine in Victorian parliamentary politics', in Dorothy Porter and Roy Porter (eds), *Doctors, Politics and Society: Historical Essays* (Amsterdam: Rodopi, 1993), pp. 81–106.
110. S. Shapin, *A Social History of Truth: Civility and Science in Seventeenth Century England* (Chicago and London: University of Chicago Press, 1994), see especially the methodological first chapter, pp. 3–41.
111. Secord, *Victorian Sensation*, pp. 405–406.
112. The mainly high church followers of John Hutchinson (1674–1737), who promoted an anti-Newtonian physico-theology based on biblical hermeneutics. See J.F.M. Clark, 'History from the ground up: bugs, political economy and God in Kirby and Spence's *Introduction to Entomology*, 1825–1856', *Isis*, 97 (2006), 28–55, 47–48.
113. Ibid., 31.
114. J.B. Morrell, 'Professors Robison and playfair, and the "Theophobia Gallica": natural philosophy, religion and politics in Edinburgh', *Notes and Records of the Royal Society of London*, 26(1) (1971), 43–63.
115. Ibid.
116. P.M. Jones, 'Living the enlightenment and the French Revolution: James Watt, Matthew Boulton, and their sons', *Historical Journal*, 42 (1999), 157–182; E. Robinson, 'An English Jacobin: James Watt, junior, 1769–1848', *Cambridge Historical Journal*, 11 (1955), 349–355.
117. Cannon, *Science in Culture*, pp. 240–244.
118. L. Goldman, 'The origins of British social science: political economy, natural science and statistics', *Historical Journal*, 26 (1983), 587–616, esp. 610, 613.
119. Jones to Whewell, 18 February 1834, ibid., 613.
120. Ibid., 614–615.

121. M. Poovey, *A History of the Modern Fact: Problems of Knowledge in the Sciences of Wealth and Society* (Chicago and London: University of Chicago Press, 1998), pp. 308–317.
122. See N. Chambers (ed.), *The Letters of Sir Joseph Banks, A Selection, 1768–1820* (London: Imperial College Press, 2000), pp. 122–125, 155–156, 163, 185.
123. Brougham to Banks, 10 December 1800, in H. Brougham, *The Life and Times of Henry Lord Brougham Written by Himself* (3 vols, Edinburgh and London: W. Blackwood and Sons, 1871), I, pp. 227–228.
124. Fox to Banks, 7 May 1802, discussed in J. Gascoigne, *Joseph Banks and the English Enlightenment: Useful Knowledge and Polite Culture* (Cambridge: Cambridge University Press, 2003), p. 21.
125. R. Drayton, *Imperial Science and Scientific Empire: Kew Gardens and the Uses of Nature, 1772–1903* (Ann Arbor, MI., 1993) [facsimile of PhD Thesis, Yale University, 1993], pp. 107–108.
126. Ibid.
127. See C.A. Bayly, *The Birth of the Modern World, 1780–1914* (Malden, MA and Oxford: Blackwell, 2004), p. 110; D.P. Miller, 'Joseph Banks, empire and "centers of calculation" in late Hanoverian London', and J. Gascoigne, 'The ordering of nature and the ordering of empire: a commentary', in D.P. Miller and P.H. Reill (eds), *Visions of Empire: Voyages, Botany and Representations of Nature* (Cambridge: Cambridge University Press, 1996), pp. 21–37, 107–116; J. Gascoigne, *Science in the Service of Empire: Joseph Banks, the British State and the Uses of Science in the Age of Revolution* (Cambridge: Cambridge University Press, 1998).
128. Drayton, *Imperial Science*, pp. 57–77.
129. Ibid., pp. 115–118.
130. C.A. Bayly, *Imperial Meridian: The British Empire and the World, 1780–1830* (London and New York: Longman, 1989), pp. 116–121.
131. Drayton, *Imperial Science*, p. 118.
132. Expanded in R. Drayton, *Nature's Government: Science, Imperial Britain, and the 'improvement' of the World* (New Haven and London: Yale University Press, 2000).
133. Drayton, *Imperial Science*, p. 104.
134. Golinski, *Science as Public Culture*, pp. 217–218.
135. Comprehensively surveyed by Thackray and Morrell, in *Gentlemen of Science: Early Years of the British Association for the Advancement of Science* (Oxford: Clarendon, 1981).
136. Drayton, *Imperial Science*, pp. 212–232.
137. A. Hume, *The Learned Societies and Printing Clubs of the United Kingdom: Being An Account of Their Respective Origin, History, Objects and Constitution* (London: Longman, Brown, Green and Longmans, 1847), preface: see pp. 3–50, 67–203.

2 The statesman

1. Naotaka Kimizuka analyses the role of the 'elder statesman' in 'Elder Statesmen and British party politics: Wellington, Lansdowne and the ministerial crises in the 1850s', *Parliamentary History*, 17(3) (1998), 355–372. Late

Victorian and Edwardian international and imperial ramifications are well served: see, for example, John S. Galbraith, 'British war aims in World War I: a commentary on "Statesmanship" ', *Journal of Imperial and Commonwealth History*, 13(1) (1984), 25–45. The preceding phase is partially explored by Harvey Mansfield, *Statesmanship and Party Government: A Study of Burke and Bolingbroke* (Chicago and London: University of Chicago Press, 1965).

2. T. Macaulay, 'The Earl of Chatham' [*Edinburgh Review*, October, 1844], in Grieve (ed.), *Critical and Historical Essays by Thomas Babington Macaulay*, I, pp. 477–478.
3. J.S. Mill, 'Considerations on Representative Government' [1861], in A.D. Lindsay (ed.), J.S. Mill, *Utilitarianism, Liberty, Representative Government* (London: J.M. Dent and Sons Ltd., 1910), Chapter XIII, p. 328; Walter Bagehot, 'Introduction to the Second Edition' [1872] prefacing *The English Constitution* (London: Kegan Paul, Trench, Trubner and Company Ltd., 1929), p. xxii. The communicating function of the statesman is more pronounced in Bagehot than in Macaulay or Mill: linked by him to the 'new suffrage' (i.e. of 1867), ibid., pp. xxi–xxvii.
4. J. Dalglish, *Eight Metaphysical Poets* (Oxford: Heinemann, 1961), pp. 81–82.
5. Thomas Hobbes, *Leviathan* [1651], in H. Morley (ed.), *Hobbes' Leviathan, Harrington's Oceana and Famous Pamphlets* (London: George Routledge and Sons, 1889); John Locke, *Two Treatises of Government* [1690], in P. Laslett (ed.) (Cambridge: Cambridge University Press, 1988).
6. See, for example, Hobbes, 'Of the public ministers of Sovereign Power', *Leviathan*, Part II, Chapter XXIII, pp. 112–115.
7. Locke, *Two Treatises*, Book I, Chapter XI, p. 154.
8. Carysfort to Lansdowne [copy], 13 November 1820, LRO 920 DER 14 box 115/1.
9. Ibid.
10. Mansfield, *Statesmanship and Party Government*, pp. 20–65, 69–70, 229–246.
11. Samuel Johnson, *A Dictionary of the English Language* (London: Knapton, 1755).
12. Adam Ferguson, *An Essay on the History of Civil Society* [1767], in Fania Oz-Salzberger (ed.) (Cambridge; Cambridge University Press, 2001), pp. 47, 133, esp. p. 173.
13. Skinner, *Reason and Rhetoric in the Philosophy of Hobbes* (Cambridge: Cambridge University Press, 1996), makes it the governing problem of Hobbesian philosophy.
14. Sir James Steuart, *An Inquiry into the Principles of Political Oeconomy* (1767), discussed in Robert Urquhart, 'The trade wind, the statesman and the system of commerce: Sir James Steuart's vision of political economy', *European Journal of the History of Economic Thought*, 33 (1996), 379–410.
15. Adam Smith, *An Inquiry into the Nature and Causes of the Wealth of Nations* [1776], in E. Cannan (ed.) (New York and Toronto: Random House, 2000), Book IV, Chapter II, p. 485.
16. Ibid., p. 841.
17. Donald Winch, *Riches and Poverty: An Intellectual History of Political Economy in Britain, 1750–1834* (Cambridge: Cambridge University Press, 1996), p. 94.
18. Smith, *Wealth of Nations*, p. 498.

19. This minor correction supports the general thrust of Winch's argument, which is to rehabilitate Smith as a political theorist: *Riches and Poverty*, pp. 90–123.
20. Edmund Burke, 'Speech on a motion for leave to bring in a Bill to repeal and alter certain Acts respecting religious opinion', 11 May 1792, in F. Willis (ed.), *The Works of the Right Honourable Edmund Burke* (6 vols, Oxford and London: Oxford University Press, 1906–07), III, p. 317.
21. G. Canning, speech at the close of the sixth day's poll, 14 October 1812, in T. Kaye, *The speeches and Public Addresses of the Rt. Hon. George Canning during the Late Election in Liverpool and on a Public Occasion in Manchester* (Liverpool: T. Kaye, London: J. Murrray, 1812), p. 19. See also the *Liverpool Poll Book* (Liverpool, 1812) in LRO H324/242/PAR.
22. W. Roscoe, *A Review of the Speeches of the Rt. Hon. George Canning, on the Late Election for Liverpool, as far as they relate to the Questions of Peace and Reform* (Liverpool and London, December 1812).
23. *Liverpool Poll Book* (1812) passim.
24. Stanley to Lansdowne, n.d. [copy] and reply, Lansdowne to Stanley, 12 September 1822, LRO 920 DER 14 box 115/1 (unfoliated).
25. Concerning this early alignment see W.D. Jones, *Lord Derby and Victorian Conservatism* (Oxford: Basil Blackwell, 1956), p. 12; discussing William Huskisson (1770–1830) to Lansdowne, 1 September 1827, in Lewis Melville (ed.), *The Huskisson Papers* (New York: R.R. Smith, 1931), p. 238, and Sir Herbert Maxwell (ed.), *The Creevey Papers* (New York: E.P. Dutton, 1904), p. 470. Jones' argument is corroborated in BL. Add. MS. 38750 ff.180–181.
26. Lansdowne to Stanley, 12 September 1822, LRO 920 DER 14 box 115/1, passim.
27. Ibid.
28. Francis Horner to Lord Webb Seymour, 4 September 1811, in L. Horner (ed.), *Memoirs and Correspondence of Francis Horner, M.P.* (2 vols, London: John Murray, 1843), II, p. 101.
29. Ibid.
30. Ibid.
31. Lyell to Ticknor, 26 December 1846, in M. Lyell (ed.), *Life, Letters and Journals of Sir Charles Lyell* (2 vols, London: John Murray, 1881), II, p. 119.
32. Lyell to his father, 16 December 1846, in ibid., II, pp. 116–117.
33. Ibid., p. 117.
34. Lyell to Ticknor, 26 December 1846, ibid., p. 119.
35. Ibid., C.C. Greville, *The Past and Present Policy of England towards Ireland* (London: E. Moxon, 1845).
36. Ibid., see Gladstone's reference to this work in W.E. Gladstone, 'The history of 1852–60 and Greville's latest Journals', *English Historical Review*, II 6 (1887), 281–302, 281–282.
37. Lyell to his father, 16 December 1846, Lyell, *Life, Letters and Journals*, II, p. 116.
38. J. Burrow, 'From Carlylean Vulcanism to sedimentary gradualism', in S. Collini, R. Whatmore and B. Young (eds), *History, Religion and Culture: British Intellectual History 1750–1950* (Cambridge: Cambridge University Press, 2000), pp. 198–223, esp. pp. 211–223.

39. D. Forbes, *The Liberal Anglican Idea of History* (Cambridge: Cambridge University Press, 1952), pp. 75–82.
40. H.H. Milman, *Quarterly Review* LXXVIII (1848), 385–386, discussed in Forbes, *The Liberal Anglican Idea of History*, p. 4.
41. Charles Lyell to his sister, 26 February 1830, Lyell, *Life, Letters and Journals*, I, p. 263. For Lyell's social intercourse with the Milmans see, for example, ibid, II, pp. 8, 29–34.
42. R. Porter, 'Charles Lyell and the principles of the history of geology', *British Journal for the History of Science*, 9 (1976), 91–103, esp. 96–97; M.J.S. Rudwick, 'Historical analogies in the geological work of Charles Lyell', *Janus*, 64 (1977), 89–107, esp. 92–93. D.R. Oldroyd claims a resemblance existed between Lyell's thought and Savigny's historical school of law: 'Historicism and the rise of historical geology', *History of Science* 17 (1979), 191–213, 227–257, see 192–193, 243–246 and 256 n.198. Lyell's direct contact with Milman in Whig environments tends to corroborate these arguments.
43. The word 'magisterial' has of course been used widely to describe the respectable, princely and didactic Reformation in contrast to the radical wing of that movement. See, for example, W.J. Torrance Kirby, *The Theology of Richard Hooker in the Context of the Magisterial Reformation* (Princeton: Princeton Seminary Press, 2000).
44. Though not by Edmund Burke. See W.J. Ashworth, 'John Herschel, George Airy, and the roaming Eye of the state', *History of Science*, 26 (1998), 151–178, esp. 155–156.
45. See ibid., 153–158; W.J. Ashworth, '"System of terror": Samuel Bentham, accountability and dockyard reform during the Napoleonic Wars', *Social History*, 23 (1998), 63–79; and more broadly, *Customs and Excise: Trade, Production and Consumption in England 1640–1845* (Oxford: Oxford University Press, 2003), especially Part V, pp. 261–315; J. Brewer, *The Sinews of Power: War, Money and the English State, 1688–1783* (London: Unwin Hyman, 1989); J. Innes, 'The domestic face of the military fiscal state: government and society in eighteenth-century Britain', in L. Stone (ed.), *An Imperial State at War: Britain from 1689 to 1815* (London: Routledge, 1993), pp. 96–127; Julian Hoppit, 'Reforming Britain's weights and measures, 1660–1824', *English Historical Review*, 108 (1998), 82–104; 'Political arithmetic in eighteenth century England', *Economic History Review*, 49 (1996), 516–540; and also Robert Poole, *Time's Alteration: Calendar Reform in Early Modern England* (London: University College London Press, 1998).
46. Ashworth, 'System of terror', 78–79.
47. I am grateful to Joanna Innes for this thought (private e-mail exchange).
48. Senior to Lansdowne, 19 June 1831, Aberystwyth, National Library of Wales, Nassau Senior Correspondence, C.164.
49. Ibid.
50. Ibid.
51. Ibid.
52. Ibid.
53. BL. Add. MS. Lans (3) 22 (P) [unfoliated].
54. In Dropmore Papers, BL. Add. MSS. 59434–5. See 59434 f.23, note on Butler's system of analogy.

55. B. Hilton, *Age of Atonement: The Influence of Evangelicalism on Social and Economic Thought, 1785–1865* (Oxford: Clarendon Press, 1991), p. 232.
56. BL. Add. MS. 59434, ff.12, 20, 22, 23–24.
57. Ibid. BL. Add. MS. 59434 f.27.
58. Hilton, *Age of Atonement*, p. 232.
59. BL. Add. MS. 59434 ff.25–26.
60. Ibid., f.28.
61. Ibid., f.27. First draft.
62. Ibid., f.4.
63. J.G.A. Pocock, *Virtue, Commerce and History* (Cambridge: Cambridge University Press, 1985), p. 285.
64. Dinwiddy, *Radicalism and Reform*, p. 29.
65. L.S. Jacyna, *Philosophic Whigs: Medicine, Science and Citizenship in Edinburgh, 1789–1848* (London, New York: Routledge, 1994), pp. 15–36.
66. Ibid., esp. p. 31. Jacyna identifies texts such as John Trenchard's *Cato's Letters* [1733] and David Fordyce's *Dialogues Concerning Education* [1745] as establishing this discourse.
67. L.G. Mitchell, *Holland House* (London: Duckworth, 1980), p. 187.
68. S. Collini, D. Winch, and J. Burrow, *That Noble Science of Politics* (Cambridge: Cambridge University Press, 1983), p. 30.
69. Holland to Shelburne, 1 September 1793, British Library, Add. MS. 51682.
70. L.G. Mitchell, *C. J. Fox and the Disintegration of the Whig Party* (London, Oxford University Press, 1971), p. 238.
71. Pocock, *Virtue, Commerce and History*, pp. 298–299.
72. Francis Horner, 17 March 1797 and 11 February 1799, in L. Horner (ed.), *Memoirs and Correspondence of Francis Horner*, I, pp. 36, 70.
73. John Clive, *Scotch Reviewers: The Edinburgh Review, 1802–1815* (London: Faber and Faber, 1957), pp. 79, 84, 106–109.
74. Pocock, *Virtue, Commerce and History*, p. 297.
75. John Campbell, *The Lives of the Lord Chancellors and Keepers of the Great Seal of England, from the Earliest Times till the Reign of Queen Victoria* (8 vols, London: John Murray, 1846–69), VIII, pp. 213–214.
76. Ibid.
77. Ibid.
78. Lansdowne to Stanley, 12 September 1822, LRO 920 DER 14 box 115/1.
79. Holland to Roscoe, 24 May 1807, LRO Roscoe Papers 2092.
80. Clive, *Scotch Reviewers*.
81. As in Collini, Winch and Burrow, *That Noble Science of Politics*.
82. Fontana, *Rethinking the Politics of Commercial Society: The Edinburgh Review, 1802–1832* (Cambridge: Cambridge University Press, 1985), p. 87, note 35.
83. Jacyna, *Philosophic Whigs*, pp. 9–49.
84. New, *The Life of Henry Brougham to 1830* (Oxford: Clarendon Press, 1961), pp. 14–16.
85. G.N. Cantor, 'Henry Brougham and the Scottish methodological tradition', *Studies in the History ad Philosophy of Science*, 2(1) (1971), 69–89.
86. W.E. Houghton (ed.), *The Wellesley Index to Victorian Periodicals: Tables of Contents and Identification of Contributors with Bibliographies of their Articles and Stories* (5 vols, London and Toronto: University of Toronto Press, 1966–79), I, pp. 416–553, esp. 430–442. The Edinburgh Review

section, including the years 1802–23, is generally taken as the authoritative guide to authorship. Compare Aspinall, *Lord Brougham and the Whig Party* (Manchester: University of Manchester, 1927), pp. 2, 5–6.

87. *Edinburgh Review*, I–V. Contributions to the *Edinburgh* were printed anonymously, but subsequent analysis corroborates Brougham's direct authorship of a large number of natural scientific articles (around 20) between the years 1802 and 1807. Several more were jointly written, and a couple have disputed authors.
88. *Edinburgh Review*, XI–XIV.
89. R. Yeo, 'An idol of the market-place: Baconianism in 19th century Britain', *History of Science*, 23 (1985), 251–298.
90. I. Loudon, 'Medical education and medical reform', in V. Nutton and R. Porter (eds), *The History of Medical Education in Britain* (Amsterdam: Rodopi, 1995), pp. 231–232.
91. J. Playfair, 'Dissertation the third, exhibiting a general view of the progress of mathematical and physical science since the revival of letters in Europe', in M. Napier (ed.), *Supplement to the Fourth, Fifth and Sixth Editions of the Encyclopaedia Britannica with Preliminary Dissertations on the History of the Sciences* (6 vols, Edinburgh: A. Constable and Company, 1824), I, pp. 458–459.
92. Cantor, 'Scottish methodological tradition', 70–79.
93. Ibid., 82–89.
94. *Edinburgh Review*, I (January 1803), 450–456.
95. Cantor, 'Scottish methodological tradition', 79–82.
96. S. Rashid, 'Dugald Stewart, '"Baconian" methodology, and political economy', *Journal of the History of Ideas*, 46 (1985), 245–257.
97. Ibid., 251–253.
98. Ibid., 85.
99. *Edinburgh Review* III (January 1804), 386–401.
100. Brougham to Loch, 20 August and 7 November 1802, 28 January 1803, in R.H.M. Buddle Atkinson and G.A. Jackson (eds), *Brougham and his Early Friends, 1798–1809* (3 vols, London: Privately printed, 1908), I, pp. 345, 363–66; II, p. 32.
101. Ibid., II, p. 32.
102. Stewart, *Henry Brougham, 1778–1868: His Public Career* (London: Bodley Head, 1985), pp.14–23, 33–36.
103. *Edinburgh Review*, III (October, 1803), 1–26, see 3.
104. *Edinburgh Review*, I (Jan. 1803), 450–456.
105. Ibid.
106. Ibid.
107. Ibid.
108. Ibid.
109. *Edinburgh Review*, XXII (January, 1808), 390.
110. Ibid., 398.
111. *Edinburgh Review*, XII (July, 1808), 399.
112. *Edinburgh Review* (September, 1814), 492.
113. *Edinburgh Review*, IV (July, 1804), 399–419.
114. Ibid., 417.
115. Ibid., 417–418.

116. Q. Skinner, *Liberty before Liberalism* (Cambridge: Cambridge University Press, 1998), pp.1–57.
117. G.N. Cantor, 'The Academy of Physics at Edinburgh, 1797–1800', *Social Studies of Science* (1975), 109–134.
118. Jacyna, *Philosophic Whigs*, p. 30.
119. Stewart, *Henry Brougham*, pp. 4–5; H. Brougham, *The Life and Times of Henry Lord Brougham Written by Himself* (3 vols, Edinburgh and London: W. Blackwood and Sons, 1871), I, pp. 18–20.
120. Jacyna, *Philosophic Whigs*, p. 30.
121. Ibid., p. 28.
122. Brougham, *Life and Times*; see also Jacyna, *Philosophic Whigs*, p. 34.
123. Brougham, *Address of Lord Brougham, President, in Opening the Congress of the National Association for Promoting Social Science, held at Glasgow on September 24th 1860* (London and Glasgow: R. Griffin and Company, 1860), p. 46.
124. *Edinburgh Review*, XII (July, 1808), 399.
125. Ibid., 400.
126. *Edinburgh Review*, III (October, 1803), 1–26.
127. Ibid., 22–26.
128. Morrell, 'Professors Robison and Playfair and the "Theophobia Gallica"', 51, 57.
129. Ibid. and see *Edinburgh Review*, III (October, 1803), 1–26.
130. *Edinburgh Review*, V (October, 1804), 97–103.
131. *Edinburgh Review*, III (October, 1803), 3.
132. *Edinburgh Review*, I (January, 1803), 426–431.
133. Yeo, 'An idol of the marketplace: Baconianism in nineteenth-century Britain', *History of Science*, 23 (1985) 251–298.
134. *Edinburgh Review*, XIII (October, 1808), 215–234.

3 Rational sociability

1. T. Macaulay, 'Lord Holland' [*Edinburgh Review*, July, 1841], in A.J. Grieve (ed.), *Critical and Historical Essays*, pp. 650–659, esp. pp. 656–658.
2. L.G. Mitchell, *Holland House* (London: Duckworth, 1980), pp. 102–109; L.S. Jacyna, *Philosophic Whigs: Medicine, Science and Citizenship in Edinburgh, 1789–1848* (London, New York: Routledge, 1994), pp. 51–77.
3. Mitchell, *Holland House*, pp. 13, 18.
4. Holland to Roscoe, 12 November and 1 December 1800, LRO Roscoe Papers, 2082–2083.
5. R.H. Shryock, 'The history of quantification in medical science', *Isis*, 52, 2 (1961), 215–237, esp. 223.
6. Holland to Roscoe, 24 May 1807, LRO Roscoe Papers, 2092.
7. Holland, *Memoirs of the Whig Party during My Time* (London: Longman, Green, Brown and Longmans, 1852), I, pp. 46–7, 55.
8. Holland, *Further Memoirs of the Whig Party with Some Miscellaneous Reminiscences, 1807–1821* (London: John Murray, 1905), pp. 371–374.
9. C.C. Gillispie (ed.), *Dictionary of Scientific Biography* (16 vols, New York: Scribner, 1970–80), V, 55.
10. Holland, *Further Memoirs*, p. 374.

11. L. Belloni, 'Anatomica plastica: III. Die Florentiner Wachsplastiken', *Ciba Symposium*, 8 (1960), 129–132.
12. Holland, *Further Memoirs*, p. 371.
13. Ibid., pp. 378–381.
14. Ibid., p. 374.
15. Belloni, 'Die Florentiner Wachsplastiken', 132.
16. Holland to Lansdowne, 9 August 1794, BL. Add. MS. 51682.
17. Ibid.
18. Lansdowne to Holland, 1 September 1794, BL. Add. MS. 51682.
19. See M. Rebecchi, 'Paolo Andreani, un viaggiatore illuminato tra il settecento e l'ottocento', *Acme*, 54(2) (2001), 143–167, esp. 151, 159, 162.
20. T. Aubert, 'Alexander Aubert, F.R.S., Astronome, 1730–1805', *Notes and Records of the Royal Society of London*, 9(1) (1951), 79–95, esp. 90.
21. Jacyna, *Philosophic Whigs*, p. 76.
22. R. Brent, *Liberal Anglican Politics: Whiggery, Religion and Reform 1830–1841* (Oxford: Clarendon, 1987), p. 112.
23. Mitchell, *Holland House*, pp. 173–174.
24. Macaulay, 'Lord Holland', pp. 656–658.
25. Mitchell, *Holland House*, pp. 175–195.
26. John Wyatt, 'George Bellas Greenough: a romantic geologist', *Archives of Natural History*, 22(1) (1995), 61–71; *Wordsworth and the Geologists* (Cambridge: Cambridge University Press, 1995), pp. 214–229.
27. R.G. Thorne, *History of Parliament: The Commons, 1790–1820* (London: Secker and Warburg for the History of Parliament Trust, 1986), vol. III, p. 172.
28. Wyatt, *Wordsworth and the Geologists*, pp. 169–193.
29. Mitchell, *Holland House*, p. 194.
30. Holland to Lansdowne, 1 September 1793, BL Add. MS. 51682.
31. Joseph Townsend, *A Journey Through Spain in the Years 1786 & 1787, with Particular Attention to the Agriculture, Manufactures, Commerce, Population, Taxes and Revenues of that Country and Remarks in Passing Through a Part of France* (3 vols, London: C. Dilly, 1791).
32. See, for example, 'Environs of Barcelona, with respect to Fossils, Agriculture and Botanical Productions', ibid. I, pp. 161–167.
33. Lansdowne to Holland, 25 July 1793, BL. Add. MS. 51682.
34. Ibid.
35. Ibid.
36. L.G. Mitchell, *Charles James Fox* (London: Penguin, 1997), pp. 178–193.
37. Sir Edmond Fitzmaurice, *Life of William, Earl of Shelburne* (2 vols, London: Macmillan and Company, 1912), II, p. 333.
38. Joseph Townsend's chief work of geology was his *Character of Moses Established for Veracity as an Historian Recording Events from the Creation to the Deluge* (2 vols, Bath, 1813 and 1815). He was also known for his *Dissertation on the Poor Laws by a Well-Wisher to Mankind* (London, 1786). See A.D. Morris, 'The Reverend Joseph Townsend MA MGS (1739–1816): physician & geologist – "Colossus of Rhodes" ', *Proceedings of the Royal Society of Medicine*, 62(5) (1969), 471–477; Hugh Torrens, 'Geological communication in the bath area in the last half of the eighteenth century', *The Practice of British Geology, 1750–1850* (Aldershot: Ashgate Variorum, 2002), pp. 217–246.

39. Charles Lyell to George Ticknor, 26 December 1846, in Sir Charles Lyell, *Life, Letters and Journals*, Mrs Lyell (ed.) (2 vols, London: John Murray, 1881), II, pp. 118–119.
40. Lansdowne to Auckland, 15 February 1816, BL. Add. MSS 34459 ff.182–183.
41. Lansdowne to Auckland, 13 September 1815, BL Add. MSS. 34459, ff.145–148.
42. Lansdowne to Auckland, 3 March 1816, BL Add. MSS. 34459, ff.186–187.
43. Ibid.
44. John Norris, *Shelburne and Reform* (London: Macmillan, 1963), pp. 84–86, 141–143, 279–281; Albert Goodwin, *The Friends of Liberty: The English Democratic Movement in the Age of the French Revolution* (Cambridge, MA: Harvard University Press, 1979), pp. 100–106; Stanhope to Shelburne [Lansdowne], 23 November 1794, Maidstone, Centre for Kentish Studies, Stanhope Papers, U1590. C56.
45. Shelburne [Lansdowne] to Grafton, 16 December 1795, Bury St. Edmunds, Suffolk Record Office, Grafton Papers 423/745.
46. Lansdowne to Auckland, 13 September 1815 and 3 March 1816, BL Add. MSS 34459 ff.145–148, 186–187.
47. See J.N. Hayes, 'Science and Brougham's society', *Annals of Science*, 20 (1964), 227–241; S. Shapin and B. Barnes, 'Science, nature and control: interpreting Mechanics' Institutes', *Social Studies of Science*, 7 (1977), 31–74; Desmond, *Politics of Evolution: Morphology, Medicine and Reform in Radical London* (Chicago and London: University of Chicago Press, 1989), pp. 30–31, 202–205, 223, 225–226, 391.
48. See, for example, William Thomas, *Philosophic Radicals: Nine Studies in Theory and Practice, 1817–1841* (Oxford: Clarendon Press, 1979), p. 56; and Austin Mitchell, *The Whigs in Opposition, 1815–1830* (Oxford: Clarendon Press, 1967), pp. 54, 125–131.
49. Jekyll to Shelburne, n.d. 1796 MS. Shelburne 52/78 [MS. Film. 2004].
50. Desmond, *Politics of Evolution*, pp. 30–31.
51. 'The London University and the Society for Diffusing Useful Knowledge', *Athenaeum* (1833) cited in Desmond, *Politics of Evolution*, p. 27.
52. Shapin and Barnes, 'Science, nature and control', 34, 65, n.9.
53. H. Brougham, *Address of Lord Brougham to the Members of the Manchester Mechanics' Institution on Tuesday 21st July 1835, with a Report of the Proceedings of the General Meeting then Held* (Manchester: Taylor and Garnett, 1835), pp. 9–29.
54. Ibid., p. 20.
55. H. Brougham, 'Speech on laying the foundation stone of the Liverpool Mechanics' Institute, 20th July 1835', *Works of Henry, Lord Brougham* (11 vols, Edinburgh: A. and C. Black, 1872–73), X, 77–89.
56. Ibid., 88–89.
57. Ibid., 85.
58. E. Burke, 'Thoughts on the cause of the present discontents' [1770], in I. Harris (ed.), *Edmund Burke: Pre-Revolutionary Writings* (Cambridge: Cambridge University Press, 1993), p. 129.
59. Ibid., p. 187.
60. Ibid., pp. 190–191.
61. Ibid.

62. Henry Grey Bennet, 'Diary of the House of Commons', 12 February 1821, HLRO, HL/PO/RO/1/129, pp. 18–19.
63. Lord John Russell, *An Essay on the History of the English Government and Constitution, From the Reign of Henry VII to the Present Time* (London: Longman, Hurst, Rees, Orme and Brown, 1821), p. 133.
64. Mitchell, *Whigs in Opposition*, p. 6.
65. Cited in ibid., p. 7.
66. See J. Bord, 'Patronage, the Lansdowne Whigs and the problem of the Liberal Centre, 1827–8', *English Historical Review*, CXVII (2002), 82–84, for an account of the kind of contest that this position opened up.
67. Bennet, 'Diary of the House of Commons', HLRO, HL/PO/RO/1/129, p. 19.
68. Morrell and Thackray, *Gentlemen of Science: Early Years of the British Association for the Advancement of Science* (Oxford: Clarendon, 1981), p. 247.
69. L. Daston, 'The ideal and reality of the republic of letters in the enlightenment', *Science in Context*, 4(2) (1991), 367–386.
70. B. Whittingham-Jones, *Liverpool's Political Clubs, 1812–30* (Liverpool, 1959); article in LRO, H/329/WHI; and *Transactions of the Historic Society of Lancashire and Cheshire*, 111 (1959), 117–138.
71. F. Furet, *Interpreting the French Revolution* (Cambridge: Cambridge University Press, 1981), pp. 173–174.
72. P. James, *'Population' Malthus: His Life and Times* (London: Routledge and Kegan Paul, 1979), p. 209
73. Brougham to Babbage, n.d.?1832, BL. Add. MS. 37187 ff.310–314.
74. J.A. Secord, *Victorian Sensation: The Extraordinary Publication, Reception, and Secret Authorship of Vestiges of the Natural History of Creation* (Chicago: University of Chicago Press, 2000), p. 416.
75. See W.J. Brock and A.J. Meadows, *The Lamp of Learning: Taylor and Francis and the Development of Science Publishing* (London: Taylor and Francis, 1998), pp. 45–47, for almanacs and calendars.
76. Hume, *The Learned Societies and Printing Clubs of the United Kingdom: Being An Account of Their Respective Origin, History, Objects and Constitution* (London: Longman, Brown, Green and Longmans, 1847), p. 31.
77. Ibid., pp. 416–418. See also N. Rupke, *Richard Owen, Victorian Naturalist* (New Haven, CT and London: Yale University Press, 1994), pp. 55–59.
78. J.A. Secord, *Controversy in Victorian Geology: The Cambrian-Silurian Dispute* (Princeton, NJ: Princeton University Press, 1986), pp. 14–24; M.J.S. Rudwick, *The Great Devonian Controversy: The Shaping of Scientific Knowledge among Gentlemanly Specialists* (Chicago: University of Chicago Press, 1985), pp. 18–30.
79. P.J. Weindling, 'Geological controversy and its historiography: the prehistory of the Geological Society of London', in L.J. Jordanova and R. Porter (eds), *Images of the Earth: Essays in the History of the Environmental Sciences* (Chalfont St Giles, British Society for the History of Science, 1979), pp. 248–271, esp. pp. 251, 255.
80. N. Rupke, *The Great Chain of History: William Buckland and the English School of Geology (1814–1849)* (Oxford: Clarendon, 1983), pp. 7, 12, 96, 99, 155.
81. M.J.S. Rudwick, *The Meaning of Fossils: Episodes in the History of Palaeontology* (Chicago: University of Chicago Press, 1985), pp. 165–168.

82. Morrell and Thackray, *Gentlemen of Science*, citing Traill's 1837 address, p. 297.
83. Ibid., pp. 302–308, 348–370.
84. The following three paragraphs are based on a comparison between the lists contained in H.B. Woodward, *The History of the Geological Society of London* (London: Longmans, Green and Company, 1907), pp. 268–285, and R.G. Thorne, *The History of Parliament: The Commons, 1790–1820* (5 vols, London: Secker and Warburg for the History of Parliament Trust, 1986), III–V, alphabetical entry by member. Woodward's lists appear to have been meticulously compiled: see Geological Society Archives, Burlington House, London, membership lists vol. F7/7. I am grateful to the archivist Mr Andrew Mussell for assistance in identifying Woodward's source.
85. See M.J.S. Rudwick, 'The foundation of the Geological Society of London: its scheme for co-operative research and its struggle for independence', *British Journal for the History of Science*, I(4) (1963), 325–355; J.B. Morrell, 'London institutions and Lyell's career: 1820–41', *British Journal for the History of Science*, 9 (1976), 132–146, esp. 135–136; Porter, *The Making of Geology: Earth Science in Britain, 1660–1815* (Cambridge: Cambridge University Press, 1977), 131, 139–140, 146–149; R. Laudan, 'Ideas and organizations in British geology: a case study in institutional history', *Isis*, 68 (1977), 527–538.
86. 'Preface', *Transactions of the Geological Society*, Ser. I, I (1811), vi–viii.
87. J. Taylor, 'On the economy of the mines of Cornwall and Devon', *Transactions of the Geological Society*, Ser.I, II (1814), 309–328.
88. H.G. Bennet, 'Some account of the island of Teneriffe', *Transactions of the Geological Society*, Ser.I, II (1814), 286–305, 303–304.
89. Lord John Russell, *The Improvement of the Law, Health, Education and Morals of the People. The Inaugural Address of Lord John Russell Delivered in St. George's Hall, Liverpool, October 11th 1858, in Connection with the Meeting of the National Association for the Promotion of Social Science* (London and Liverpool, ?1859), pp. 10–11.
90. Ibid.
91. Lord Lansdowne, *Speech of the Marquis of Lansdowne in the House of Lords, June 3rd 1818, on Moving for Certain Information Relative to the State of the Prisons in the United Kingdom* (London, 1818), pp. 13–14.
92. Lord Palmerston, *Speech of Lord Viscount Palmerston, Secretary of State for Foreign Affairs, to the Electors of Tiverton on 31st July, 1847* (2nd edn, London, 1847), p. 20.
93. Burke in the Commons, 28 May and 16 June 1794; 'The collected works of Edmund Burke', *The Speeches of the Right Honourable Edmund Burke on the Impeachment of Warren Hastings to which is Added a Selection of Burke's Epistolary Correspondence* (8 vols, in Bohn's Standard Library, London: George Bell and Sons, 1889), VII [I of the Speeches and Correspondence], 467, and VIII [II], 438. See S. Rashid's discussion in 'Dugald Stewart, Baconian methodology, and political economy', *Journal of the History of Ideas*, 46 (1985), 245–257.
94. The argument revises S.F. Cannon's thesis that Baconianism had little to do with the actual practice of nineteenth-century science: R. Yeo, 'An idol of the marketplace: Baconianism in nineteenth-century Britain',

History of Science, 23 (1985), 252; S.F. Cannon, *Science in Culture: The Early Victorian Period* (Folkestone, Dawson, 1978), pp. 58–59. The phenomenon of Baconian apologetics is further explored in A. Perez-Ramos, 'Bacon's Legacy' in M. Peltonen (ed.), *The Cambridge Companion to Bacon* (Cambridge: Cambridge University Press, 1996), esp. pp. 321–324; Golinski, *Science as Public Culture: Chemistry and Enlightenment in Britain, 1760–1820* (Cambridge: Cambridge University Press, 1992), pp. 4, 19–24, 85, 104.

95. Yeo, 'An idol of the market place', 251–298.
96. Yeo, 'An idol of the market place', 284–287.
97. J.G.A. Pocock, *Virtue, Commerce and History* (Cambridge: Cambridge University Press, 1985), pp. 190, 291, 310; Winch, *Riches and Poverty: An Intellectual History of Political Economy in Britain, 1750–1834* (Cambridge: Cambridge University Press, 1996), esp. Chapter 11, on Romantic anti-Malthusianism, pp. 288–322.
98. Pocock, *Virtue, Commerce and History*, p. 291.
99. Pocock, *Virtue, Commerce and History*, pp. 291–292.
100. John Money, 'Joseph Priestley in cultural context: Philosophic spectacle, popular belief and popular politics in eighteenth century Birmingham', *Enlightenment and Dissent*, 7 (1988), 57–81, and 8 (1989), 69–89, discussed in Golinski, *Science as Public Culture*, p. 104.
101. H. Brougham, *Practical Observations upon the Education of the People, Addressed to the Working Classes and their Employers* (London: Longman and Company, 1825), notes 3 and 5.
102. T.B. Macaulay, *Review of the Works of Francis Bacon, Lord Chancellor of* England edited by Basil Montagu, the *Edinburgh Review* (July, 1837), *Critical and Historical Essays*, II, 290–398, esp. 385–387.
103. Ibid., 377.
104. See, for example, David Brewster's attack on Bacon's reputation in the *Edinburgh Review* (1830). Also see Yeo, 'An idol of the market place', 266.
105. L. Goldman, 'The Social Science Association, 1857–1886: a context for mid-Victorian Liberalism', *English Historical Review* (January, 1986), no. 398, 95–134.
106. M. Napier, *Lord Bacon and Sir Walter Raleigh* (Cambridge, 1853), pp. 36–8.
107. D. Stewart, 'Preliminary dissertation' or 'Dissertation first: exhibiting a general view of the progress of metaphysical, ethical and political philosophy since the revival of letters in Europe', in M. Napier (ed.), *Encyclopaedia Britannica or Dictionary of the Arts, Sciences and General Literature* (7th edn, Edinburgh: A. Constable and Company, 1842), I.
108. Ibid., p. 40.
109. Ibid., p. 36.
110. Ibid., p. 37.
111. Ibid.
112. W.S. Landor, *Imaginary Conversations of Literary Men and Statesmen* (3 vols, 2nd edn, London: Colburn, 1826), II, p. 96.
113. Landor's politics are difficult to characterise, but he was a close friend of Southey and Coleridge into the post-Napoleonic period and was dismissive of Fox's historical writings: D. Eastwood, 'Robert Southey and the meanings of patriotism', *Journal of British Studies*, 31(3) (1992), 265–287, 274n;

J.R. Dinwiddy, 'Charles Fox as historian', *Historical Journal*, 12, 1 (1969), 23–34, 31n.

114. W. Blake, *Jerusalem: The Emanation of the Giant Albion* [Facsimile], in M.D. Paley (ed.) (London: William Blake Trust with the assistance of the Getty Grant Foundation, 1991), v.50.
115. S.T. Coleridge, *On the Constitution of Church and State* (London, 1830) in J. Colmer (ed.), *The Collected Works of Samuel Taylor Coleridge* (16 vols, London: Routledge and Kegan Paul, 1976), X, p. 62. Henceforth *Church and State*.
116. Ibid., p. 13.
117. Ibid. For Vichian links and parallels see G. Whalley, 'Coleridge and Vico', *Giambattista Vico: An International Symposium*, G. Tagliacozzo and H.V. White (eds) (Baltimore: Johns Hopkins Press, 1969), pp. 225–244; L. Pompa, *Vico: A Study of the 'New Science'* (2nd edn, Cambridge: Cambridge University Press, 1990), pp. 87–103; M. Lilla, *G.B. Vico: The Making of an Anti-Modern* (Cambridge MA and London: Harvard University Press, 1994), pp. 47–48, 139.
118. Horner, personal journal, 10 May 1801, in Bourne and Taylor (eds), *Horner Papers: Selections from the Letters and Miscellaneous Writings of Francis Horner, MP, 1795–1817* (Edinburgh: Edinburgh University Press, 1994), p. 155.
119. Horner, Paper dated June 1803, ibid., p. 288.
120. See, for example, his review of Lord King's 'Thoughts on the restrictions of payments in Specie at the banks of England and Ireland', *Edinburgh Review*, II (July, 1803), 402–421, reproduced in F.W. Fetter (ed.), *The Economic Writings of Francis Horner in the Edinburgh Review, 1802–6* (London: London School of Economics and Political Science, 1957), pp. 77–95.
121. Horner, 23 November 1800, in Bourne and Taylor (eds), *Horner Papers*, p. 129.
122. J.C. Robertson, 'A Bacon-facing generation: Scottish philosophy in the early nineteenth century', *Journal of the History of Philosophy*, 14(1) (1976), 37–49.

4 Liberality

1. J.H. Newman, *Apologia Pro Vita Sua: Being a History of his Religious Opinions* (London: Longmans, Green and Company, 1904), Note 'A', 'Liberalism'.
2. J.E. Cookson, *The Friends of Peace: Anti-War Liberalism in England 1793–1815* (Cambridge: Cambridge University Press, 1982), pp. 2–3.
3. Ibid., p. 3.
4. See A. Kriegel, 'Liberty and Whiggery in early nineteenth-century England', *Journal of Modern History*, 52(2) (1980), 253–278, esp. 259, citing Mackintosh to Holland, 23 February 1805, BL Add. MS 51653 f.7.
5. Bedford to Grenville, 9 February 1817, in R. Russell (ed.), *The Early Correspondence of Lord John Russell 1805–40* (2 vols, London: T.F. Unwin, 1913), I, p. 187.

6. P. Mandler, *Aristocratic Government in the Age of Reform: Whigs and Liberals, 1830–1852* (Oxford: Clarendon, 1990).
7. Kriegel, 'Liberty and Whiggery', 259.
8. J.C.D. Clark (ed.), E. Burke, *Reflections on the Revolution in France* [1790] (Stanford, CA: Stanford University Press, 2001), p. 242.
9. Ibid.
10. S. Johnson, *A Dictionary of the English Language*. See Clark (ed.), *Reflections*, p. 242 n. 302; and J.C.D. Clark, *English Society 1660–1832: Religion, Ideology and Politics during the Ancient Regime* (2nd edn, Cambridge: Cambridge University Press, 2000), p. 6.
11. D. Winch, *Riches and Poverty: An Intellectual History of Political Economy in Britain, 1750–1834* (Cambridge: Cambridge University Press, 1996), pp. 166, 175–176.
12. Clark, *English Society*, pp. 6–7.
13. Clark, *English Society*, p. 6; Cookson, *Friends of Peace*, p. 3. For a further example see R. Wright, *An Apology for Dr Michael Servetus Including an Account of his Life, Persecution, Writings and Opinions, Being Designed to Eradicate Bigotry and Uncharitableness & to Promote Liberality of Sentiment among Christians* (Wisbech: F.B. Wright, 1806), vi–vii and pp. 383–384, on removal of Catholic disabilities.
14. T. Finch, *Address to the Poor of Northrepps in the County of Norfolk, on an Occasion of Private Beneficence to them, & Delivered at the said Parish Church on Thursday the 22nd January 1795; Containing some Civil and Religious Principles suitable to the Times* (Norwich: Bacon, 1795), pp. 8–9.
15. S. Glasse, *National Liberality and National Reform Recommended: A Sermon Preached in the Parish Church of St George, Bloomsbury, on Sunday February 4th 1798* (London: Printed by request 1798).
16. See N. Aston, 'Glasse, Samuel (1734–1812)', *Oxford Dictionary of National Biography* (61 vols, Oxford: Oxford University Press, 2004), XXII.
17. Matt. 12:21; Glasse, *National Liberality*, pp. 3–4.
18. Ibid., p. 11.
19. Ibid., p. 14.
20. G. Preston, *The Principles of Modern Liberality and Fanaticism Inconsistent with the Simplicity of Gospel Truth. A Sermon Preached in the church of St. Peter, Colchester, at the Visitation of the Reverend Joseph Jefferson, M.A, Archdeacon of Colchester on Tuesday May 18th 1819* (Colchester, 1819).
21. Ibid., p. 25.
22. Ibid., pp. 10–11.
23. A. Bagnall, *Antiquated Scrupulosity Contrasted with Modern Liberality Occasioned by Henry Gally Knight's 'Foreign and Domestic View of the Catholic Question'* (London, 1829), pp. 5–6, 43.
24. William Sharpe, *Considerations on Modern Liberality and on Civil Disobedience: Two Assize Sermons Preached at the Assizes holden for the County of Somerset in the Year 1830* (London, 1830), pp. 1–5.
25. Daniel Sandford, Bishop of Edinburgh, diary 2 May 1828, 'Remains', as cited by William Sharpe, *Considerations*, p. 25. See J. Sandford (ed.), *Remains of the late Right Reverend Daniel Sandford, Including Extracts from his Diary and Correspondence, and a Selection from his Unpublished Sermons, with a Memoir by the Reverend John Sandford* (Edinburgh, 2 vols, 1830).

26. Lansdowne to Stanley, 12 September 1822, LRO 920 DER 14 box 115/1 unfoliated.
27. J.G.A. Pocock, *Virtue, Commerce and History* (Cambridge: Cambridge University Press, 1985), pp. 254–255; L. G. Schwoerer, 'William, Lord Russell, the Making of a Martyr', 1683–1983', *Journal of British Studies*, 24(1) (1985), 41–71, esp. 51 and 61–68.
28. R. Brent, *Liberal Anglican Politics: Whiggery, Religion and Reform 1830–1841* (Oxford: Clarendon, 1987), p. 54.
29. Lord John Russell, 'Ministerial Plan of Parliamentary Reform', 1 March 1831, in J. Russell, *Selections from Speeches of Earl Russell 1817 to 1841 and from Despatches 1859 to 1865* (2 vols, London: Longmans, Green, 1870), I, p. 334.
30. Brent, *Liberal Anglican Politics*, pp. 52–53.
31. Ibid., citing Russell, *Speeches*, I, p. 236.
32. Livy, 'The history of Rome', VI.40–42, in R.M. Ogilvie (ed.) and B. Radice (trans.), *Livy: Rome and Italy, Books VI–X of the History of Rome from its Foundation* (Harmondsworth: Penguin, 1982), pp. 90–96.
33. Brent, *Liberal Anglican Politics*, pp. 52–57.
34. Russell, *Speeches*, I, p. 333.
35. N. Machiavelli, *Discourses on the First Ten Books of Titus Livy* [c. 1517], in B. Crick (ed.), L.J. Walker and B. Richardson (trans.) (Harmondsworth: Penguin, 1971), I–51, p. 234, and see the prudential argument in I–32, 'Neither a Republic nor a Prince should put off conferring benefits on people until danger is at hand', pp. 188–189.
36. Russell, *Speeches*, I, p. 238.
37. N. Machiavelli, *The Prince* [1513], in P. Bondanella (ed. and trans.), M. Musa (trans.) (Oxford: Oxford University Press, 1984), XVI, 'On Generosity and Miserliness', pp. 53–55. John Burrow points out that Russell excerpted *The Prince* in *A History of English Government*: J.W. Burrow, *A Liberal Descent. Victorian Historians and the English Past* (Cambridge: Cambridge University Press, 1981), p. 29 n.71.
38. Brent, *Liberal Anglican Politics*, p. 52–55.
39. Burrow, *A Liberal Descent*, p. 29.
40. Ibid., p. 33.
41. J. Burrow, S. Collini, and D. Winch, *That Noble Science of Politics* (Cambridge: Cambridge University Press, 1983), p. 16; J.W. Burrow, *Whigs and Liberals: Continuity and Change in English Political Thought* (Oxford: Clarendon, 1988), pp. 11, 28–29.
42. Burrow, Collini, and Winch, *Noble Science of Politics*, pp. 188–189.
43. M. Viroli, *Machiavelli* (Oxford and New York: Oxford University Press, 1998), p. 148.
44. T.B. Macaulay, 'Machiavelli', *Edinburgh Review*, 90 (1827), reproduced in F.C. Montague (ed.), *Critical and Historical Essays Contributed to the Edinburgh Review* (3 vols, London: Methuen and Company, 1903), I, pp. 61–111.
45. Ibid.
46. Reverend Sydney Smith as cited by W. Bagehot in his review, 'The first Edinburgh Reviewers' [1855], in R.H. Hutton (ed.), *Literary Studies*, (2vols, London: Longman, Green and Company), I, pp. 15–16.See also the letter in Smith (ed.), Reverend Sydney Smith to J.A. Murray, 4 June 1843, *Selected Letters of Sydney Smith*, pp. 312–313.

47. Cited by William Roscoe (1753–1831) in his attempt to encourage Lansdowne to take a copy of the first part of his *Monandrian Plants of the Order Scitaminae* (Liverpool, 1824), letter from Roscoe to Lansdowne, 4 April 1824; Lansdowne's reply politely declining, 14 April 1824, Liverpool, Liverpool Record Office (LRO) 920 ROS 2358–9.
48. Lansdowne also served on the Council of the Geological Society, 1816–17: Woodward, *The History of the Geological Society of London* (London: Longmans, Green and Company, 1907), pp. 268–285. Morris Berman, *Social Change and Scientific Organisation: The Royal Institution, 1799–1844* (London: Heinemann Educational, 1978), pp. 109–111, 120–123. Patricia James notices the accession of a significant number of Whigs to the Geological Society in 1813, including Lansdowne, John Whishaw and the Duke of Devonshire: *'Population' Malthus*, p. 209.
49. Bagehot, *Literary Studies*, I, p. 16.
50. See, for example, Morrell and Thackray, *Gentlemen of Science: Early Years of the British Association for the Advancement of Science* (Oxford: Clarendon, 1981); Roy Porter, 'Gentleman and geology: the emergence of a scientific career, 1660–1920', *Historical Journal*, 21(4) (December 1978), 809–836, especially Porter's evocation of the 'knights of the hammer', 815–825.
51. Hilton, 'The politics of anatomy and an anatomy of politics, c.1825–1850', in Collini, Whatmore, and Young (eds), *History, Religion and Culture*, p. 193.
52. Ibid.
53. See, for example, Desmond, *Politics of Evolution*, esp. pp. 12–13, 374–378; L.S. Jacyna, 'Immanence or Transcendence: theories of life and organisation in Britain, 1790–1835', *Isis*, 74(3) (1983), esp. 321–329; Morrell, 'Professors Robison and Playfair and the "Theophobia Gallica": natural philosophy, religion and politics in Edinburgh', *Notes and Records of the Royal Society of London*, 26(1) (1971), 43–63.
54. Earl Fitzwilliam to Rev. Henry Zouch, 5 June 1792, discussed in Smith, *Whig principles and Party Politics: Earl Fitzwilliam and the Whig Party, 1748–1833* (Manchester: Manchester University Press, 1975), p. 138.
55. Lord John Russell, *An Essay on the History of the English Government and Constitution, from the Reign of Henry VII to the Present Time* (London, 1821), pp. 2, 17; and the division of public power under 'Gothic monarchies', Ibid., p. 53.
56. Lord Althorp to Henry Brougham, 5 July 1835, BL. Althorp Papers, H2 [unfoliated].
57. Jacyna, *Philosophic Whigs*, pp.52–73.
58. J. Allen, *Inquiry into the Rise and Growth of the Royal Prerogative in England* (London, 1830), pp. 27–28. See Jacyna's discussion in *Philosophic Whigs*, p. 76.
59. Ibid.
60. Allen, *Royal Prerogative*, p. 29.
61. Ibid., p. 4.
62. Ibid., pp. 4–5.
63. Althorp to Brougham, 24 June 1835, BL Althorp Papers H1 [unfoliated].
64. Lord John Russell, *Essays on the History of the Christian Religion* (2nd edn, London: Longmans, Green and Company, 1873), pp. 110–111.
65. Ibid., pp. 300–301.
66. Brent, *Liberal Anglican Politics*.

67. Lansdowne to Auckland, 14 April 1817, BL. Add. MS. 34459, ff.192–193.
68. Lansdowne to Auckland, 16 July 1818, BL. Add. MS. 34459 ff.267–269; T. Macaulay, 'Ranke's History of the Popes', *Critical and Historical Essays* (2 vols, London, 1946), II, pp. 60–62 [October, 1840].
69. T. Macaulay, 'Francis Bacon', *Critical and Historical Essays*, II, p. 364 [July, 1837].
70. B. Hilton, *Age of Atonement: The Influence of Evangelicalism on Social and Economic Thought, 1785–1865* (Oxford: Clarendon Press, 1991), pp. 238–239.
71. Russell, *English Government and Constitution*, p. 46.
72. Allen, *Royal Prerogative*, p. 15.
73. Ibid., pp. 15–16, 164.
74. Russell, *Essays on the History of the Christian Religion*, p. 308.
75. Brent, *Liberal Anglican Politics*, pp. 58–64. Brent catches the tyranny but neglects the ontological issue.
76. J. Prest, *Lord John Russell* (London: Macmillan, 1972), p. 322; E.R. Norman, *Anti-Catholicism in Victorian England* (London: Allen and Unwin, 1968), p. 57.
77. N. Gash, *Reaction and Reconstruction in English Politics, 1832–1852* (Oxford: Clarendon Press, 1965), p. 114.
78. Lord John Russell, *Papal Aggression: Substance of a Speech of the Rt. Hon. Lord John Russell Delivered in the House of Commons, February 7 1851* (London: Longman, Brown, Green and Longmans, 1851), pp. 40–41.
79. Ibid., pp. 17–18, 24.
80. Ibid., pp. 9, 15.
81. Hilton, *Age of Atonement*, p. 350. See W.E. Gladstone, *Ecclesiastical Titles Assumption Bill: Speech in the House of Commons, 25th March, 1851* (London, 1851); P. Butler, *Gladstone: Church, State and Tractarianism, A Study of his Religious Ideas and Attitudes, 1809–1859* (Oxford: Clarendon, 1982), pp. 142–144; Norman, *Anti-Catholicism in Victorian England*, pp. 75–77.
82. John David Yule, 'The impact of science on British religious thought in the second quarter of the nineteenth century', PhD thesis (University of Cambridge, 1976), 187–235. H. Brougham's *Dialogues on Instinct* (1839) are noted briefly for their pretension, ibid. 193.
83. R.J. Richards, 'Instinct and intelligence in British natural theology: some contributions to Darwin's Theory of the Evolution of Behavior', *Journal of the History of Biology*, 14(2) (1981), 193–229, esp. 209–214.
84. Althorp to Brougham, 24 June 1835, BL Althorp Papers, H1 [unfoliated].
85. Brougham to Althorp, ?16 June 1835, BL. Althorp Papers, H4 [unfoliated].
86. Ibid.
87. Althorp to Brougham, 24 June 1835, BL. Althorp Papers, H1 [unfoliated].
88. Ibid.
89. Brougham, *Dialogues on Instinct with Analytic View of the Researches on Fossil Osteology* [1839], in Bell and Brougham (eds) *Paley's Natural Theology*, V (repr. London, 1844), p.62.
90. Ibid., p. 61.
91. Ibid., p. 65.
92. H. Brougham, 'Balance of power', *Edinburgh Review*, I (January 1803), reproduced in 'Dissertations – historical and political', *Works of Henry Lord Brougham* (11 vols, Edinburgh: A. and C. Black, 1872–73), VIII, pp. 2–3.

93. Ibid., 11; See D. Hume, 'Of the balance of power', in K. Haakonssen (ed.), *Political Essays* (Cambridge and New York: Cambridge University Press, 1994), pp. 154–160. The cross-reference is explicit.
94. Ibid., pp. 157–158.
95. Brougham, 'Balance of power', 11–12.
96. Ibid., p. 12.
97. Brougham, *Dialogues on Instinct*, p. 64.
98. Althorp to Brougham, 31 May 1836, BL. Althorp Papers H2 [unfoliated].
99. Althorp to Brougham, 17 February 1836, BL. Althorp Papers, H2 [unfoliated].
100. Ibid., annotation; and Brougham, *Dialogues on Instinct*, First Dialogue.
101. Ibid.
102. Althorp to Brougham, 31 May 1836, BL. Althorp Papers, H2 [unfoliated].
103. Ibid.
104. Ibid.
105. Ibid.
106. Ibid.
107. Ibid.
108. Brougham, *Dialogues on Instinct*, p. 64. The words put into 'Althorp's' mouth as a note of caution nevertheless describe a 'delightful' position that 'Brougham' goes on to defend.
109. Brougham, *Dialogues on Instinct*, p. 163.
110. Althorp to Brougham, 24 September 1835, BL. Althorp Papers H2 [unfoliated].
111. H. Brougham, *A Discourse of Natural Theology Showing the Nature of the Evidence, and the Advantages of the Study* in Bell and Brougham (eds) *Paley's Natural Theology* (5 vols, London: Charles Knight, 1835–39)—, I, pp. 212–213: also citing Job, chs. 38–41.
112. Yule, 'The impact of science on British religious thought', 200.

5 The Georgic tradition

1. See D. Eastwood, 'Robert Southey and the intellectual origins of Romantic Conservatism', *English Historical Review*, CIV (1989), 308–331, esp. 311–320; John Stevenson, 'William Cobbett: patriot or Briton?' *Transactions of the Royal Historical Society*, 6th Series, 6 (1996), 123–136, esp. 125–126, 128.
2. W.D. Rubinstein, *Elites and the Wealthy in Modern British Society* (Brighton: Harvester, 1987), pp. 129–132.
3. S.T. Coleridge, *On the Constitution of the Church and State*, in Colmer (ed.), *The Collected Works of Samuel Taylor Coleridge*, (16 vols, London: Routledge and Kegan Paul, 1976), X, 68, 218.
4. J.G.A. Pocock, *Virtue, Commerce and History* (Cambridge: Cambridge University Press, 1985), p. 218.
5. Ibid.
6. The pattern of landholding at the beginning of the period has been explored by G.E. Mingay, *English Landed Society in the Eighteenth Century* (London: Routledge and Paul, 1963).

7. The economic history of the period has been extensively surveyed: J.H. Clapham, *An Economic History of Modern Britain* (3 vols, Cambridge: Cambridge University Press, 1926–38), I and II; also see W.D. Rubinstein, *Elites and the Wealthy*; F.M.L. Thompson, *English Landed Society in the Nineteenth Century* (London: Routledge and Kegan Paul; Toronto: University of Toronto Press, 1963); D. Cannadine, *Aspects of Aristocracy: Grandeur and Decline in Modern Britain* (London: Penguin, 1995).
8. Lansdowne Estate, 'Reports, accounts and rentals of the estate of the Marquess of Lansdowne in County Kerry, 1864–73', Dublin, National Library of Ireland, N818/P1020 [Microfilm], pp. 1–5 [including retrospective table], 52–53.
9. E.A. Wasson, *Whig Renaissance: Lord Althorp and the Whig Party, 1782–1845* (New York and London: Garland, 1987), pp. 16–17.
10. Smith, *Whig Principles and Party Politics*, p. 31.
11. Ibid., xiii.
12. Ibid., p. 29.
13. Clark (ed.), Burke, *Reflections on the Revolution in France*, p. 31.
14. Virgil, *The Georgics*, L.P. Wilkinson (ed. and trans.) (Harmondsworth: Penguin, 1982); *The Eclogues* trans. G. Lee (Harmondsworth: Penguin, 1984).
15. A good description of this canon can be found in D.I. Allsobrook, 'The Georgic model of middle-class education' in *Schools for the Shires: The Reform of Middle-Class Education in Mid-Victorian England* (Manchester: Manchester University Press, 1988), pp. 28–57, esp. pp. 30–31. See also Walter Harte, *Essays on Husbandry* (London and Bath: W. Frederick, 1764), p. 18.
16. L.P. Wilkinson, *The 'Georgics' of Virgil: A Critical Survey* (London: Cambridge University Press, 1969), pp. 223–269, esp. pp. 299–313.
17. M.R.D. Foot (ed.), *The Gladstone Diaries* (22 vols, Oxford: Clarendon Press, 1968), II, pp. 445–446.
18. L.G. Mitchell, *Charles James Fox* (London: Penguin, 1997), p. 186.
19. A. Hemingway, *Landscape Imagery and Urban Culture in Early Nineteenth-Century Britain* (Cambridge and New York: Cambridge University Press, 1992), pp. 293–297.
20. Ibid., p. 130.
21. See D.H. Solkin, *Richard Wilson: The Landscape of Reaction* (London: Tate Gallery, 1982), pp. 22–34, 40.
22. J. Barrell, *The Dark Side of the Landscape: The Rural Poor in English Painting, 1730–1840* (Cambridge: Cambridge University Press, 1980), pp. 35–88, esp. pp. 37–41; see also Ann Bermingham, *Landscape and Ideology: The English Rustic Tradition, 1740–1860* (London: Thames and Hudson, 1987), pp. 5, 204.
23. Ibid., pp. 80–81.
24. Ibid., p. 77.
25. Lord John Russell (ed.), *Correspondence of John, Fourth Duke of Bedford, Selected from the Originals at Woburn Abbey* (3 vols, London: Longman, 1842–46), I, p. liii.
26. Attributed to the sixth Duke of Bedford, Ibid., lv–lvi.
27. Ibid., lxxxii.

28. Lord John Russell, 'Preface', in H. Cradock, *The Calendar of Nature, Or The Seasons of England, Edited with A Preface by the Right Honourable Lord John Russell* (London, n.d., c. 1850), i–iii.
29. Ibid.
30. See Prest, *Lord John Russell*, pp. 3–6. See the Bedford–Young correspondence in BL Add. MS. 35128–9, the primary source for this paragraph.
31. E. Clarke, 'Agriculture and the House of Russell', *Journal of the Royal Agricultural Society of England*, Ser.iii I (1891), 123–145, esp. 133. This piece is a retrospective tribute in honour of the ninth Duke of Bedford.
32. Bedford to Young, 28 March 1802, BL. Add. MS. 35128 ff.431–2.
33. Ibid.
34. Bedford to Young, 28 April 1802, BL. Add. MS. 35128 ff.447–448; Prest, *Lord John Russell*, p. 5.
35. Bedford to Young, 5 July 1802, BL. Add. MS. 35128 f.473; Bedford to Young, 29 May 1803; and see the 'comparative trial' mentioned in Bedford to Young, 28 June 1804 BL. Add. MS. 35129 ff. 50, 140.
36. Bedford to Young, 9 June 1806, BL. Add. MS. 35129 ff.342–343.
37. Bedford to Young, 2 December 1806, BL. Add. MS. 35129 ff. 361–362.
38. Bedford to Young, 31 July 1804, BL. Add. MS. 35128 ff.182–3.
39. Ibid.
40. Golinski, *Science as Public Culture*, pp. 191–192, 217–218.
41. See E.A. Wasson, 'The Third Earl Spencer and agriculture, 1818–1845', *Agricultural History Review*, 26 (1978), 89–99; R.A.C. Parker, 'Coke of Norfolk and the Agrarian Revolution', *Economic History Review*, 2nd series, 8 (1955–56), 156–166.
42. Creevey to Ord, 14 April 1827, in H. Maxwell (ed.), *The Creevey Papers: A Selection from the Correspondence and Diaries of the late Thomas Creevey, MP* (London: John Murray, 1912), p. 453. See also Brougham to Creevey, 5 February 1820, Ibid., p. 297; Creevey to Ord on 'the marvellous man', 3 January 1838, ibid., p. 673.
43. Reminiscences in Clarke, 'Agriculture and the House of Russell'.
44. Ibid., 136. Clarke does not mention the Whig politics of these figures. I am most grateful for the opportunity to verify the original to Mr Phillip Sheppy, honorary librarian to the Royal Agricultural Society of England [RASE], RASE Library, Stoneleigh Park, Warks. Engraving, Inventory no. A4.1.
45. N.B. Penny, 'The Whig cult of Fox in early nineteenth-century sculpture', *Past and Present*, 70 (1976), 94–105, esp. 96–99.
46. Ibid., 99; and Clarke, 'Agriculture and the House of Russell', 133.
47. Bedford to Lord John Russell, 21 March 1838, cited in Prest, *Lord John Russell*, p. 5.
48. Ibid.
49. *Hortus Gramineus Woburnensis OR an Account of the Results of experiments on the Produce and Nutritive Qualities of Different Grasses and other Plants used as the Food of the more Valuable Domestic Animals Instituted by John, Duke of Bedford* (London, 1824), 'Advertisement'.
50. H. Davy, *Elements of Agricultural Chemistry in A Course of Lectures for the Board of Agriculture* (London: Longman, 1813), 'Advertisement'.

51. Andy Hector and Rowan Hooper, 'Darwin and the first ecological experiment', *Science*, 295 (5555), Issue of 25 January 2002, 639–640.
52. 'Origin of Darwin's gardening reference revealed', *The Independent*, 25 January 2002, 11. As Hector and Hooper acknowledge, the identification was actually made nearly thirty years ago by R.C. Stauffer, in his edition of *Charles Darwin's Natural Selection* (London: Cambridge University Press, 1975), pp. 227–251.
53. Davy, *Elements of Agricultural Chemistry*, pp. 26–27, p. 322; cited on the title page of *Hortus Gramineus Woburnensis*.
54. *Hortus*, v–vi.
55. Ibid., vi.
56. Ibid., vii.
57. *Hortus*, 'Appendix II', pp. 422–423.
58. Ibid., p. 428.
59. *Hortus*, vi. Losses in the post-war context of the second edition (1824).
60. See D. Knight, 'Agriculture and Chemistry in Britain Around 1800', *Annals of Science*, 33 (1976), 189, 196.
61. Berman, *Social Change and Scientific Organisation*, p. 61. In the caesura, Berman argues that Davy had previously shown little interest in agriculture.
62. Ibid., p. 68.
63. Davy, *Elements of Agricultural Chemistry*, p. 6.
64. Ibid., pp. 322–323.
65. Ibid.
66. Ibid.
67. Ibid.
68. Ibid., p. 24.
69. Berman, *Social Change and Scientific Organisation*, p. 70.
70. Davy, *Elements of Agricultural Chemistry*, pp. 312–313.
71. Ibid.
72. Ibid., pp. 322–323.
73. *Hortus*, Frontispiece.
74. Ibid., p. 15.
75. Ibid., pp. 322–323.
76. Ibid., pp. 24–25.
77. *Edinburgh Review*, XXII (January, 1814), 251–281.
78. Ibid., 251.
79. Ibid., 279–281.
80. *Quarterly Review*, XI (April–July, 1814), 318–332.
81. Ibid., pp. 318–320.
82. R. Siegfried and R.H. Dott, Jr. (eds), *Humphry Davy on Geology: The 1805 Lectures for the General Audience* (Madison, WI and London: University of Wisconsin Press, 1980), pp. 3–9.
83. Ibid., p. 3.
84. Ibid., p. 7.
85. Ibid., p. 6.
86. Ibid., p. 4.
87. Ibid. See 'Lecture Ten', esp. pp. 136–139.
88. Joseph Jekyll to Shelburne (the first Marquess of Lansdowne), MS. Shelburne.52.MS Film 2004 n.78 n.d. 1796.

89. Ibid.
90. B. Hilton, *Age of Atonement: The Influence of Evangelicalism on Social and Economic Thought, 1785–1865* (Oxford: Clarendon Press, 1991), p. 380.
91. Ibid. and G. Himmelfarb, *The Idea of Poverty: England in the Early Industrial Age* (London: Faber, 1985), pp. 100–147, citation p. 133.
92. Mary Poovey, *A History of the Modern Fact; Problems of Knowledge in the Sciences of Wealth and Society* (Chicago and London: University of Chicago Press, 1998), pp. 278–295.
93. Discussed in Hilton, *Age of Atonement*, p. 380.
94. Drayton, *Imperial Science and Scientific Empire: Kew Gardens and the Uses of Nature, 1772–1903* (Ann Arbor, MI., 1993) [facsimile of PhD Thesis, Yale University, 1993], p. 217.
95. Gascoigne, *Joseph Banks and the English Enlightenment: Useful Knowledge and Polite Culture* (Cambridge: Cambridge University Press, 2003), pp. 14–15, 203–216; 'Joseph Banks and his abiding legacy', *London Papers in Australian Studies*, No.2 (London, 2001), 1–12.
96. Stewart, *Henry Brougham: His Public Career*, p. 4, for the best-documented example of the well-worn tale about Scotsmen and rifles; and Thomas Love Peacock, *Crotchet Castle*, R. Wright (ed.) (Harmondsworth: Penguin, 1969), p. 135.
97. See J.F. Burke, *British Husbandry: Exhibiting the Farming Practice in Various Parts of the United Kingdom* (3 vols, London: Baldwin and Cradock, 1834–40), especially I; cf. C. Knight, *An Address to the Labourers: On the Subject of Destroying Machinery* (London: Charles Knight, 1830).
98. Brougham, *The Life and Times of Henry Lord Brougham Written by Himself* (3 vols, Edinburgh and London: W. Blackwood and Sons, 1871), I: especially the first chapter, in which Brougham constructs his genealogy.
99. In Fontana, *Rethinking the Politics of Commercial Society*; A.C. Chitnis, *The Scottish Enlightenment and early Victorian English Society* (London: Croom Helm, 1985); Winch, *That Noble Science of Politics* (Cambridge: Cambridge University Press, 1983), pp. 25–61.
100. Jacyna, *Philosophic Whigs*, p. 13.
101. M. McNeil, *Under the Banner of Science: Erasmus Darwin and his Age* (Manchester: Manchester University Press, 1987), pp. 169–183.
102. Jacyna, *Philosophic Whigs*, p. 32.
103. D.R. Dean, *James Hutton and the History of Geology* (Ithaca and London: Cornell University Press, 1992), p. 264.
104. Dugald Stewart, *Collected Works*, Sir William Hamilton (ed.) (11 vols, Edinburgh: Thomas Constable and Company; London: Hamilton, Adams and Company, 1854–60), IV, 8. See Virgil, *Eclogues*, pp. 58–60.
105. Stewart, *Collected Works*, ibid.; and Virgil, *Eclogues*, p. 146.
106. Walter Harte, *Essays on Husbandry* (London and Bath: W. Frederick, 1764), p. 209.
107. Adam Smith, *Lectures on Rhetoric & Belles-Lettres* [1762–63], in J.C. Bryce (ed.), *Works and Correspondence of Adam Smith*, (Oxford: Clarendon Press, 1983), IV, 142–145.
108. Ibid.
109. Stewart, *Collected Works*, III, pp. 160–161.

110. Ibid.
111. Ibid. and compare 'Outlines of Moral Philosophy' in *Collected Works*, II, p. 6.
112. Stewart, *Collected Works*, vol. III, p. 167.
113. See *Eclogues*, p. 57 [iv. 34].
114. Stewart, *Collected Works*, vol. III, p. 167.
115. Ibid. and p. 168.
116. Clark, *Reflections*, p. 93.
117. Bourne and Taylor, *The Horner Papers*, pp. 185–191.
118. Ibid., p. 185.
119. Ibid., pp. 186, 189 n.10.
120. Poovey, *History of the Modern Fact*, pp. 283, 290.
121. Bourne and Taylor, *Horner Papers*, pp. 186–187.
122. Ibid., p. 190 n.16. See B. Dobrée, *The Letters of Philip Dormer Stanhope, 4th Earl of Chesterfield* (London: Eyre and Spottiswoode, 1932), Letter 158.
123. T. Malthus, *An Essay on the Principle of Population* [1798], in Anthony Flew (ed.) (Harmondsworth: Penguin, 1979), pp. 81–86.
124. Ibid., p. 83.
125. Ibid.
126. Harte, *Essays on Husbandry*, p. 89.
127. Ibid.
128. Ibid., p. 210.
129. Ibid., p. 11.
130. Ibid., p. 41.
131. Stewart, *Collected Works*, IX, 286–288.
132. Harte, *Essays on Husbandry*, pp. 26–28.
133. G.W. Kitchen (ed.), F. Bacon, *The Advancement of Learning* (London: J.M. Dent and Sons Ltd., 1861), Book II, p. 154.
134. Dugald Stewart, *Preliminary Dissertation to the Encyclopaedia Britannica or Dictionary of the Arts, Sciences and General Literature*, Macvey Napier (ed.) (21 vols, Edinburgh: A. and C. Black, 1842), I.
135. Ibid., 32–36.
136. Ibid., 36.
137. Ibid., 38.
138. Ibid.
139. Stewart, *Collected Works*, IX, 362.
140. Ibid.
141. Stewart, 'Preliminary Dissertation', *Encyclopaedia Britannica*, I, pp. 36–37.
142. Ibid.
143. Ibid., 39–40.
144. Ibid.
145. Bacon [ed. Kitchen], *Advancement of Learning*, pp. 154–155.
146. E.F. Biagini, 'Neo-roman liberalism: "republican" values and British liberalism, ca. 1860–1875', *History of European Ideas*, 29 (2003), 55–72.
147. G.K. Roberts, 'The establishment of the Royal College of Chemistry: an investigation of the social context of early-Victorian chemistry', *Historical Studies in the Physical Sciences*, 7 (1976), 437–485. See also N.G. Coley, G.K. Roberts, and C.A. Russell, *Chemists by Profession: The origins and rise of the Royal Institute of Chemistry* (Milton Keynes: Open University Press for the Royal Institute of Chemistry, 1977), pp. 76–93; and more briefly,

on the Royal College of Agriculture, Allsobrook, *Schools for the Shires*, pp. 37–38.

148. See N. Goddard, *Harvests of Change: The Royal Agricultural Society of England, 183–1988* (London: Quiller Press, 1988), pp. 2–3.
149. Ibid., 99.
150. E.A. Wasson, 'The Third Earl Spencer and agriculture, 1818-45',*Agricultural History Review*, 26 (1978), 89–99.
151. Ducie to Peel, and Gardner (note) to Peel, 25 October 1844, BL. Add. MS. 40553 f.22. The *Prospectus Proposed for Establishing a College of Chemistry for Promoting the Science and its Application to Agriculture, Arts, Manufactures and Medicine* (London: Privately printed, 1844) was enclosed.
152. Hilton, 'Politics of anatomy' in Collini, Whatmore and Young, *History, Religion and Culture*, p. 183.
153. Ibid.
154. Roberts, 'Royal College of Chemistry', 460–461.
155. Ibid., 465.
156. Ibid.
157. J. Bell (ed.), *Pharmaceutical Journal and Transactions*, VI (1846–47), 155.
158. Roberts, 'Royal College of Chemistry', 465 n. 79.
159. Ibid., 461.
160. *Pharmaceutical Journal*, 157–159.
161. Ibid., 157–158.
162. Ibid., 158.
163. Ibid., 157.
164. The latter term was coined by Howard Becker in *Outsiders: Studies in the Sociology of Deviance* (London: Free Press of Glencoe, 1963) and is appropriate to rule formation in a mature capitalist state. In contrast, moral adventurers such as Gardner and Bullock navigated the conventions of pre-democratic patronage networks.
165. Roberts, 'Royal College of Chemistry', 466–468.
166. BL. Add. MS. 40553 ff.23–30, 'Prospectus', p. 2.
167. Ibid., pp. 6, 11.
168. Ibid., p. 13.
169. Ibid., p. 8.
170. Ibid., pp. 8–9.
171. Ibid., p. 13.
172. Ibid.
173. Peel to Liebig, 25 March 1843, BL. Add. MS. 40526 ff.78–79.
174. Peel to Ducie, 28 October 1844, BL. Add. MS. 40553 f.31.
175. Ibid.

Conclusion

1. J. Bord, 'Patronage, the Lansdowne Whigs and the problem of the liberal centre, 1827–8', *English Historical Review*, CXVII (2002), 78–93, 83. See also J.C.D. Clark, *English Society 1660–1832: Religion, Ideology and Politics during the Ancient Regime* (2nd edn, Cambridge: Cambridge University Press, 2000), pp. 5–8.

2. Isaac Kramnick, 'Eighteenth-century science and radical social theory: the case of Joseph Priestley's scientific liberalism', *Journal of British Studies*, 25 (January 1986), 1–30.
3. Ibid., esp. p. 10, and M. Bentley (ed.), *Public and Private Doctrine: Essays in British History Presented to Maurice Cowling* (Cambridge: Cambridge University Press, 1993). See also J.E. Cookson, *The Friends of Peace: Anti-War Liberalism in England 1793–1815* (Cambridge: Cambridge University Press, 1982), pp. 2–4.
4. Especially in J.G.A. Pocock's synthetic essay, 'The varieties of Whiggism from Exclusion to Reform: a history of ideology and discourse', *Virtue, Commerce and History* (Cambridge: Cambridge University Press, 1985), pp. 215–310.
5. R. O'Connor, 'Mammoths and Maggots: Byron and the Geology of Cuvier', *Romanticism*, 5 (1999), 26–42, esp. 26, 29.
6. The path-breaking work is J.G.A. Pocock, *The Machiavellian Moment: Florentine Political Thought and the Atlantic Republican Tradition* (Princeton, NJ and London: 2nd edn, Princeton University Press, 2003). See L.S. Jacyna, *Philosophic Whigs: Medicine, Science and Citizenship in Edinburgh, 1789–1848* (London, New York: Routledge, 1994), esp. pp. 15–29.
7. *Manchester Guardian*, 16 October 1824; see A. Mitchell, *The Whigs in Opposition, 1815–1830* (Oxford: Clarendon Press, 1967), p. 56.
8. *Ipswich Journal*, 17 February 1821, discussed in Mitchell, *Whigs in Opposition*, pp. 55–56.
9. H. Brougham, *A Discourse of Natural Theology Showing the Nature of the Evidence and the Advantages of the Study* (3rd edn, London: Charles Knight, 1835), pp. 150–151.
10. See, for example, T. Porter, *The Rise of Statistical Thinking: 1820–1900* (Princeton, NJ and Guildford: Princeton University Press, 1986); *Trust in Numbers: The Pursuit of Objectivity in Science and Public Life* (Princeton, NJ and Chichester: Princeton University Press, 1995), esp. pp. 51–53, and the general model in pp. 194–231; M. Poovey, *A History of the Modern Fact: Problems of Knowledge in the Sciences of Wealth and Society* (Chicago and London: University of Chicago Press, 1998), esp. pp. 308–317.
11. The emblematic aphorism is 'Words are deeds': Q. Skinner, in J. Tully (ed.), *Meaning and Context, Quentin Skinner and His Critics* (Princeton, NJ: Princeton University Press, 1988), frontispiece; see also J. Tully, 'The pen is a mighty sword: Quentin Skinner's analysis of politics', ibid., pp. 8, 23; Q. Skinner, '"Social meaning" and the explanation of social action', ibid., pp. 79–80; 'On meaning and speech-acts', ibid., p. 260; J. Farr, 'Understanding conceptual change politically', in T. Ball, J. Farr, and R.L. Hanson (eds), *Political Innovation and Conceptual Change* (Cambridge: Cambridge University Press, 1989), pp. 24–31; H.F. Pitkin, 'Representation', ibid., p. 132; Pocock, 'The state of the art', *Virtue, Commerce and History*, pp. 1–37.
12. Jacyna, *Philosophic Whigs*, p. 48.
13. Ibid., p. 6.
14. Q. Skinner in Tully (ed.), *Meaning and Context*, pp. 32, 36, 68–78.
15. Thus Skinner begins his investigation of *Reason and Rhetoric in the Philosophy of Hobbes* (Cambridge: Cambridge University Press, 1996) by examining the rhetorical and oratorical manuals of Tudor schools, ibid., pp. 19–65.

16. Skinner, *Liberty before Liberalism* (Cambridge: Cambridge University Press, 1998), pp. 101–109; Pocock, *Virtue, Commerce and History*, pp. 59–71, 258–260.
17. Skinner, *Liberty before Liberalism*, pp. 103–104; Pocock, *Virtue, Commerce and History*, p. 34.
18. Skinner in Tully (ed.), *Meaning and Context*, pp. 109–110.
19. L. Namier, *The Structure of Politics at the Accession of George III* (2nd edn, London: Macmillan and Company, 1957), p. xi.
20. Ibid., p. x.
21. *Meaning and Context*, pp. 108–111, 127–128.
22. Q. Skinner, 'The principles and practice of opposition: the case of Bolingbroke versus Walpole', in N. McKendrick (ed.), *Historical Perspectives; Essays in Honour of J.H. Plumb* (London: Europa, 1974), pp. 93–128, citation p. 128. See also S.M. Lee, 'A new language in Politicks; George Canning and the Idea of Opposition, 1801–1807', *History*, 83, 271 (1998), 474–477.
23. Bord, 'Patronage, the Lansdowne Whigs and the problem of the liberal centre', 78–93.
24. C. Babbage, memorandum to Lord John Russell, December 1846, BL. Add. MS. 37183, ff.347–357.
25. Ibid., BL. Add. MS. 37183, f.355.
26. Ibid., f.356.
27. Ernst Mayr, 'When is historiography Whiggish?', *Journal of the History of Ideas*, 51 (1990), 301–309, citation 302.
28. Ibid. This approach is typified in Mayr's *The Growth of Biological Thought* (Cambridge MA: Belknap Press, 1982).
29. A. Desmond, *The Politics of Evolution: Morphology, Medicine and Reform in Radical London* (Chicago and London: University of Chicago Press, 1989), p. 23.
30. P. Mandler, 'Aristocratic styles in the age of Reform I & II', *Aristocratic Government in the Age of Reform: Whigs and Liberals, 1830–1852* (Oxford: Clarendon, 1990), pp. 44–120.
31. G.L. Geison, 'Research schools and new directions in the historiography of science', *Osiris*, 8 (1993), 227–238, esp. 236–238, and 'Scientific change, emerging specialties, and research schools', *History of Science*, 10 (1981), 20–40; J.B. Morrell, 'The chemist breeders: the research schools of Liebig and Thomson', *Ambix*, 19 (1972), 1–46; J.S. Fruton, *Contrasts in Scientific Style: Research Groups in the Chemical and Biochemical Sciences* (Philadelphia: American Philosophical Society, 1990). See the symposium devoted to 'Style in Science', comprising the whole issue of *Science in Context*, 4 (2) (1991), 223–447.
32. L. Daston and M. Otte, 'Introduction', *Science in Context*, 4 (2) (1991), 223.
33. Ibid., 227.
34. The application is defended in N. Reingold, 'The peculiarities of the Americans or are there national styles in the sciences?', *Science in Context*, 4 (2) (1991), 347–366, definition 349: but see Anna Wessely, 'Transposing "Style" from the history of art to the history of science', ibid., 265–278, esp. 270–271. Clearly, the objection has much less force in cases where pursuit of a 'national' style was a self-conscious goal. See Anne Harrington, 'Interwar "German" psychobiology: between nationalism and the irrational', ibid., 429–447.

35. Geison, 'Research schools and new directions', 227, 236–238.
36. Surveyed by Jan Golinski, 'The theory of practice and the practice of theory: sociological approaches in the history of science', *Isis*, 81 (1990), 492–505. See also Ian Hacking, *The Social Construction of What?* (Cambridge MA, and London: Harvard University Press, 1999), pp. 63–99.
37. For a recent example see J.D. Mollon, 'The origins of the concept of interference', *Philosophical Transactions of the Royal Society of London A* (2002), 360, 807–819: particularly the discussion of Thomas Young (1773–1829) and trichromacy, 816–818. Desmond is optimistic in thinking that 'few historians see their task any more as reconstructing a rational lineage of ideas through time', *The Politics of Evolution*, p. 21.
38. James A. Secord, *Victorian Sensation: The Extraordinary Publication, Reception, and Secret Authorship of Vestiges of the Natural History of Creation* (Chicago: University of Chicago Press, 2000).
39. Aggressive impulses: Namier, *The Structure of Politics at the Accession of George III*, p. 1.
40. A useful discussion of this topic is to be found in Jean Chalaby, 'Beyond the prison-house of language: discourse as a sociological concept', *British Journal of Sociology*, 47 (4) (1996), 684–697.
41. Desmond, *Politics of Evolution*, p. 21–22.
42. For example, J.C.D. Clark, 'A general theory of party, opposition and government, 1688–1832', *Historical Journal*, 23 (2) (1980), 295–325.
43. C. Darwin, *On the Origin of Species By Means of Natural Selection* [1859], in J. Carroll (ed.) (Peterborough, Ont: Broadview Press Ltd., 2003), Ch. I, pp. 99–131.
44. See, for example, C. Kenneth Waters, 'The arguments in the *Origin of Species*', J. Hodge and G. Radick (eds), *Cambridge Companion to Darwin*, (Cambridge: Cambridge University Press, 2003), pp. 119–127.
45. The concept of political selection has been used before to denote a process of naturalistic selection in political structures. In contrast, we are using it here to describe a variant of artificial selection. See M. Wohlgemuth, 'Evolutionary approaches to politics', *Kyklos*, 55 (2002), 2, 223–246; A. Farkas, *State Learning and International Change* (Ann Arbor, MI: University of Michigan Press, 1988), pp. 72, 170.
46. Darwin, *Origin*, pp. 107, 114, 116.
47. Cited in Bagehot, *Literary Studies* (2 vols, London: Longman, Green and Company), I, p. 15. See also N.C. Smith, *Selected Letters of Sydney Smith* (London, Oxford University Press, 1956), pp. 312–313.
48. Ibid., p. 146 (in Ch. IV).
49. See S.J. Gould, 'More things in heaven and earth', reproduced in S. Rose (ed.), *The Richness of Life: The Essential Stephen Jay Gould* (London and New York: W.W. Norton and Company, 2007), pp. 444–466, esp. p. 457. Gould argued that cultural evolution was Lamarckian in style, ibid., p. 615.
50. Pocock, 'Varieties of Whiggism', *Virtue, Commerce and History*, passim.
51. See, for example, M. Bentley, *The Climax of Liberal Politics: British Liberalism in Theory and Practice, 1868–1918* (London: Edward Arnold, 1987).
52. Darwin, *Origin*, pp. 116–121.
53. Goldman, 'The origins of British social science: political economy, natural science and statistics', *Historical Journal*, 26 (1983).

54. Darwin, *Origin*, p. 116.
55. Ibid., p. 119.
56. Ibid. (Ch. IV), p. 144–146.
57. I am of course taking for granted the role of natural selection in producing the physical and anthropological basis of human behaviour. But this is beyond or perhaps behind the purview of the political historian.
58. J.W. Burrow, *Whigs and Liberals: Continuity and Change in English Political Thought* (Oxford: Clarendon, 1988).
59. Mandler, *Aristocratic Government*.
60. R. Brent, *Liberal Anglican Politics: Whiggery, Religion and Reform 1830–1841* (Oxford: Clarendon, 1987). See also B. Hilton, 'Whiggery, religion and social reform: the case of Lord Morpeth', *Historical Journal*, 37 (1994)', 829–835.
61. Brent, *Liberal Anglican Politics*, p. xii.
62. Clark, *English Society*, pp. 5–8.

Bibliography

Manuscripts and abbreviations

Aberystwyth

National Library of Wales, Nassau Senior Papers.

Bury St. Edmunds

Suffolk Record Office, Grafton Papers.

Dublin

National Library of Ireland, Lansdowne Estate Papers, Larcom Papers.

Edinburgh

National Library of Scotland, Dugald Stewart Papers.

Liverpool

Liverpool Record Office (LRO), Central Library, Derby Papers, Roscoe Papers.

London

British Library (BL), Additional Manuscripts Collection (Add. MS: Lansdowne Papers, Auckland Papers, Holland Papers, Huskisson Papers, Babbage Papers, Dropmore Papers, Arthur Young Papers, Peel Papers, Althorp Papers).
——House of Lords Record Office (HLRO): Henry Grey Bennet Diary.

Maidstone

Centre for Kentish Studies, Stanhope Papers.

Oxford

Bodleian Library, Shelburne Papers (microfilm).

Winchester

Hampshire Record Office, Tierney Papers.

Printed primary sources

Documents

Historical Manuscripts Commission, Fortescue Manuscripts (Dropmore) (London, 1912), VIII.

Journals and periodicals

The Anti-Jacobin Review and Magazine, or Monthly Political and Literary Censor.
Edinburgh Review or Critical Journal.
Quarterly Review.

Other works

Allen, J., *Inquiry into the Rise and Growth of the Royal Prerogative in England* (London, 1830).

Bacon, F., *The Advancement of Learning*, G.W. Kitchen (ed.) (London: J.M. Dent and Sons, 1861).

Bagehot, W., *Literary Studies by the Late Walter Bagehot M.A. and Fellow of University College London with Prefatory Memoir*, R.H. Hutton (ed.) (2 vols, 2nd edn, London: Longmans, Green and Company, 1879), I.

——*The English Constitution* (London: Kegan Paul, Trench, Trubner and Company, 1929).

Bagnall, A., *Antiquated Scrupulosity Contrasted with Modern Liberality Occasioned by Henry Gally Knight's 'Foreign and Domestic View of the Catholic Question'* (London, 1829).

Banks, J., *The Letters of Sir Joseph Banks, A Selection, 1768–1820*, N. Chambers (ed.) (London: Imperial College Press, 2000).

Bell, C. and Brougham, H., *Paley's Natural Theology with Illustrative Notes* (5 vols, London: Charles Knight, 1835–39), I–V.

Bell, J. (ed.) *Pharmaceutical Journal and Transactions*, VI (1846–47).

Bennet, H.G., 'Some account of the island of Teneriffe', *Transactions of the Geological Society*, Ser.I, II (1814), 286–305.

Bentham, J., *The Correspondence of Jeremy Bentham*, J.R. Dinwiddy (ed.) (12 vols, London: Athlone Press; and vols 6–12, Oxford: Clarendon Press, 1988), VI–VII.

Blake, W., *Jerusalem: The Emanation of the Giant Albion*, M.D. Paley (ed.) (London: William Blake Trust with the assistance of the Getty Grant Foundation, 1991).

Brougham, H., *Practical Observations upon the Education of the People, Addressed to the Working Classes and their Employers* (London: Longman and Company, 1825).

——*A Discourse of the Objects, Advantages and Pleasures of Science* (2nd edn, London: Baldwin and Cradock, 1827).

——*A Discourse of Natural Theology showing the Nature of the Evidence, and the Advantages of the Study* (3rd edn, London: Charles Knight, 1835).

——*Address of Lord Brougham to the Members of the Manchester Mechanics Institution on Tuesday 21st July 1835, With a Report of the Proceedings of the General Meeting Then Held* (Manchester: Taylor and Garnett, 1835).

——, *Dialogues on Instinct with Analytic View of the Researches on Fossil Osteology* [1839] in Bell and Brougham above, IV–V (repr. London: C. Knight and Company, 1844).

——, *Address of Lord Brougham, President, in Opening the Congress of the National Association for Promoting Social Science, held at Glasgow on September 24th 1860* (London and Glasgow: R. Griffin and Company, 1860).
——, *The Life and Times of Henry Lord Brougham Written by Himself* (3 vols, Edinburgh and London: W. Blackwood and Sons, 1871), I.
——, *Works of Henry Lord Brougham* (11 vols, Edinburgh: A. and C. Black, 1872–73), VIII, X.
Buddle Atkinson, R.H.M. and G.A. Jackson (eds), *Brougham and his Early Friends, 1798–1809* (3 vols, London: privately printed, 1908), I–II.
Burke, E., 'The collected works of Edmund Burke', *The Speeches of the Right Honourable Edmund Burke on the Impeachment of Warren Hastings: To which is Added a Selection of Burke's Epistolary Correspondence* (2 supp. vols, London: George Bell and Sons, 1889), I–II.
——, *The Works of the Right Honourable Edmund Burke*, F. Willis (ed.) (6 vols, Oxford and London: Oxford University Press, 1906–07), III.
——, *Edmund Burke: Pre-Revolutionary Writings*, I. Harris (ed.) (Cambridge: Cambridge University Press, 1996).
——, *Reflections on the Revolution in France*, J.C.D. Clark (ed.) (Stanford CA: Stanford University Press, 2001).
Burke, J.F., *British Husbandry: Exhibiting the Farming Practice in Various Parts of the United Kingdom* (3 vols, London: Baldwin and Cradock, 1834–40).
Campbell, J., *The Lives of the Lord Chancellors and Keepers of the Great Seal of England, from the Earliest Times till the Reign of Queen Victoria* (8 vols, London: John Murray, 1846–69), VIII.
Canning, G., *The speeches and Public Addresses of the Rt. Hon. George Canning during the Late Election in Liverpool and on a Public Occasion in Manchester*, T. Kaye (ed.) (Liverpool: T. Kaye, London: J. Murray, 1812).
Chesterfield, Philip Dormer Stanhope, 4th Earl – *The Letters of Philip Dormer Stanhope, 4th Earl of Chesterfield* ed. B. Dobrée (London: Eyre and Spottiswoode, 1932).
Cobbett, W., *Cobbett's Parliamentary History of England to the Year 1803* (36 vols, London: T.C. Hansard for Longman, Hurst, Rees, Orme and Brown, 1806–20), XXXII.
Coleridge, S.T., *On the Constitution of Church and State*, J. Colmer (ed.), *The Collected Works of Samuel Taylor Coleridge* (16 vols, London: Routledge and Kegan Paul, 1976), X.
Cradock, H., *The Calendar of Nature, or the Seasons of England, Edited with a Preface by the Right Honourable Lord John Russell* (London, n.d., c.1850).
Creevey, T., *The Creevey Papers*, H. Maxwell (ed.) (New York: E.P. Dutton, 1904).
——, *The Creevey Papers: A Selection from the Correspondence and Diaries of the Late Thomas Creevey, MP*, H. Maxwell (ed.) (London: John Murray, 1912).
Darwin, C., *On the Origin of Species by Means of Natural Selection* [1859], J. Carroll (ed.) (Peterborough, Ont: Broadview Press Ltd., 2003), Ch. I, pp. 99–131.
Davy, H., *Humphry Davy on Geology: The 1805 Lectures for the General Audience*, R.H. Dott Jr, and R. Siegfried (ed.) (Madison WI, and London: University of Wisconsin Press, 1980).
——, *Elements of Agricultural Chemistry in a Course of Lectures for the Board of Agriculture* (London: Longman, 1813).

Ferguson, A., *An Essay on the History of Civil Society*, F. Oz-Salzberger (ed.) (Cambridge: Cambridge University Press, 2001).
Finch, T., *Address to the Poor of Northrepps in the County of Norfolk, on an Occasion of Private Beneficence to them, & Delivered at the said Parish Church on Thursday the 22nd January 1795; Containing some Civil and Religious Principles Suitable to the Times* (Norwich, 1795).
Fox, C.J., *Memorials and Correspondence of Charles James Fox*, J. Russell (ed.) (4 vols, London: R. Bentley, 1853–57).
Gardner, J., *Prospectus Proposed for Establishing a College of Chemistry for Promoting the Science and its Application to Agriculture, Arts, Manufactures and Medicine* (London: Privately printed, 1844).
Gladstone, W.E. 'The history of 1852–60 and Greville's latest journals', *English Historical Review*, II(6) (1887), 281–302.
——, *The Gladstone Diaries*, M.R.D. Foot (ed.) (22 vols, Oxford: Clarendon Press, 1968), II.
Glasse, S., *National Liberality and National Reform Recommended: A Sermon Preached in the Parish Church of St George, Bloomsbury on Sunday February 4th 1798* (London: Printed by request, 1798).
Grenville, W., *Essay on the Supposed Advantages of a Sinking Fund* (London: John Murray, 1828).
Greville, C.C. *The Past and Present Policy of England towards Ireland* (London: E. Moxon, 1845).
Greville, R.F., *The Diaries of Colonel the Hon. Robert Fulke Greville, Equerry to his Majesty King George III*, F.M. Bladon (ed.) (London: Bodley Head, 1930).
Hansard, T.C. (ed.), *Parliamentary Debates from the Year 1803 to the Present Time* (41 vols, London: Longman, Hurst, Orme, Rees and Brown, 1812–20), VII, 879–899; IX, 89–99, 1007–1015; XI, 841–844; XV, 348.
Harte, W., *Essays on Husbandry* (London and Bath: W. Frederick, 1764).
Hobbes, T., *Leviathan*, H. Morley (ed.) (London: George Routledge and Sons, 1889).
Holland, Henry Richard, *Memoirs of the Whig Party during My Time*, H.E. Holland (ed.) (2 vols, London: Longman, Green, Brown and Longmans, 1852–53), I.
——, *Further Memoirs of the Whig Party with Some Miscellaneous Reminiscences, 1807–1821*, Lord Stavordale (ed.) (London: John Murray, 1905).
Horner, F., *Memoirs and Correspondence of Francis Horner, M.P.*, L. Horner (ed.) (2 vols, London: John Murray, 1843), II.
——, *Memoirs and Correspondence of Francis Horner, M.P.*, L. Horner (ed.) (2 vols, 2nd edn, London: John Murray, 1853), I–II.
——, *The Economic Writings of Francis Horner in the Edinburgh Review, 1802–6*, F.W. Fetter (ed.) (London: London School of Economics and Political Science, 1957).
——, *The Horner Papers; Selections from the Letters and Miscellaneous Writings of Francis Horner, MP*, K. Bourne and W.B. Taylor (ed.) (Edinburgh: Edinburgh University Press, 1994).
Hume, A., *The Learned Societies and Printing Clubs of the United Kingdom: Being an Account of their Respective Origin, History, Objects and Constitution* (London: Longman, Brown, Green and Longmans, 1847).
Hume, D., *Political Essays*, K. Haakonssen (ed.) (Cambridge and New York: Cambridge University Press, 1994).
Huskisson, W., *The Huskisson Papers*, L. Melville (ed.) (New York: R.R. Smith, 1931).

Johnson, S., *A Dictionary of the English Language: In which the Words are Deduced from their Originals and Illustrated in their Different Significations by Examples from the Best Writers: To which are Prefixed, a History of the Language and an English Grammar* (London: Knapton, 1755).
Knight, C., *An Address to the Labourers: On the Subject of Destroying Machinery* (London: Charles Knight, 1830).
Landor, Walter Savage, *Imaginary Conversations of Literary Men and Statesmen* (3 vols, 2nd edn, London: Colburn, 1826).
Lansdowne, Henry Petty-Fitzmaurice, 3rd Marquess – [Petty, Lord Henry] *Statement of a Plan of Finance Proposed to Parliament in the Year 1807* (London, 1807).
——, *Speech of the Right Honourable Lord Henry Petty Delivered in the Committee of Finance* (London, 1807).
——, *Speech of the Marquis of Lansdowne in the House of Lords, June 3rd 1818, on moving for certain Information Relative to the State of the Prisons in the United Kingdom* (London, 1818).
——, *The Speech of the Marquis of Lansdown on Foreign Commerce* (London, 1821).
——, *The Speech of Lord Lansdowne on the Spanish Colonies* (London, 1824).
Liebig, J. von, *Familiar Letters on Chemistry, in its relations to Physiology, Dietetics, Agriculture, Commerce, and Political Economy* [3rd edn, London, 1851] in *The Development of Chemistry, 1789–1914* (10 vols, facsimile ed., London: Routledge, 1998), VI.
Livy, *Livy: Rome and Italy, Books VI–X of The History of Rome from its Foundation* translated and edited by R.M. Ogilvie and B. Radice (Harmondsworth: Penguin, 1982).
Locke, J., *Two Treatises of Government*, P. Laslett (ed.) (Cambridge: Cambridge University Press, 1988).
Lyell, C. *Life, Letters and Journals*, K.M. Lyell (ed.) (2 vols, London: John Murray, 1881), II.
Macaulay, T.B., *Critical and Historical Essays Contributed to the Edinburgh Review*, F.C. Montague (ed.) (3 vols, London: Methuen and Company, 1903), I.
——, *Critical and Historical Essays by Thomas Babington Macaulay*, A.J. Grieve (ed.) (2 vols, London: Dent, 1907), I–II.
Machiavelli, N., *Discourses on the First Ten Books of Titus Livy* [c.1517] B. Crick (ed.), trans. L.J. Walker and B. Richardson (Harmondsworth: Penguin, 1971).
——, *The Prince* [1513] edited and translated by P. Bondanella, trans. M. Musa (Oxford: Oxford University Press, 1984).
Malthus, T., *An Essay on The Principle of Population* [1798], A. Flew (ed.) (Harmondsworth: Penguin, 1979).
Mill, J.S., *Utilitarianism, Liberty, Representative Government*, A.D. Lindsay (ed.) (London: J.M. Dent and Sons Ltd., 1910).
Napier, M., *Lord Bacon and Sir Walter Raleigh* (Cambridge, 1853).
Palmerston, Henry John Temple, 3rd Viscount – *Speech of Lord Viscount Palmerston, Secretary of State for Foreign Affairs to the Electors of Tiverton on 31st July, 1847* (2nd edn, London, 1847).
Peacock, T. Love, *Crotchet Castle*, R. Wright (ed.) (Harmondsworth, Penguin, 1969).
Playfair, J., 'Dissertation the third, exhibiting a general view of the progress of mathematical and physical science since the revival of letters in Europe', M. Napier (ed.), *Supplement to the fourth, Fifth and Sixth Editions of the*

Encyclopaedia Britannica or Dictionary of the Arts, Sciences and General Literature With Preliminary Dissertations on the History of the Sciences (6 vols, 7th edn, Edinburgh: A. Constable and Company, 1842), I.

Preston, G., *The Principles of Modern Liberality and Fanaticism Inconsistent with the Simplicity of Gospel Truth. A Sermon preached in the church of St. Peter, Colchester at the Visitation of the Reverend Joseph Jefferson, M.A, Archdeacon of Colchester on Tuesday May 18th 1819* (Colchester, 1819).

Ricardo, D., *Works and Correspondence of David Ricardo*, P. Sraffa and M.H. Dobb (ed.) (11 vols, Cambridge, 1962), IV, VIII.

Roscoe, W., *A Review of the Speeches of the Rt. Hon. George Canning, on the Late Election for Liverpool as far as they relate to The Questions of Peace and Reform* (Liverpool and London, 1812).

——, *Monandrian Plants of the Order Scitaminae, Chiefly Drawn from Living Specimens at the Botanic Garden at Liverpool* (2 vols, Liverpool, 1824–28), I–II.

Russell, John, 1st Earl, *An Essay on the History of the English Government and Constitution, From the Reign of Henry VII to the Present Time* (London: Longman, Hurst, Rees, Orme and Brown, 1821).

——, *Correspondence of John, Fourth Duke of Bedford, Selected from the Originals at Woburn Abbey* (3 vols, London: Longman, 1842–46), I.

——, *Papal Aggression: Substance of a Speech of the Rt. Hon. Lord John Russell Delivered in the House of Commons, February 7 1851* (London: Longman, Brown, Green and Longmans, 1851).

——, *The Improvement of the Law, Health, Education and Morals of the People. The Inaugural Address of Lord John Russell Delivered in St. George's Hall, Liverpool, October 11th 1858, in Connection with the Meeting of the National Association for the Promotion of Social Science* (London and Liverpool, c.1859).

——, *Selections from Speeches of Earl Russell 1817 to 1841 and from Despatches 1859 to 1865* (2 vols, London: Longmans, Green, 1870), I.

——, *Essays on the History of the Christian Religion* (2nd edn, London: Longmans, Green and Company, 1873).

——, *The Early Correspondence of Lord John Russell 1805–40*, R. Russell (ed.) (2 vols, London: T.F. Unwin, 1913).

Sandford, J., *Remains of the Late Right Reverend Daniel Sandford, Including Extracts from his Diary and Correspondence, and a Selection from his Unpublished Sermons, with a Memoir by the Reverend John Sandford* (2 vols, Edinburgh, 1830), I–II.

Sharpe, W., *Considerations on Modern Liberality and on Civil Disobedience: Two Assize Sermons Preached at the Assizes holden for the County of Somerset in the Year 1830* (London, 1830).

Sinclair, G., *Hortus Gramineus Woburnensis OR an Account of the Results of Experiments on the Produce and Nutritive Qualities of Different Grasses and other Plants used as the Food of the more Valuable Domestic Animals Instituted by John, Duke of Bedford* (London, 1824).

Smith, A., *Works and Correspondence of Adam Smith*, J.C. Bryce (ed.) (Oxford: Clarendon Press, 1983) IV.

——, *An Inquiry into the Nature and Causes of the Wealth of Nations*, E. Cannan (ed.) (New York and Toronto: Random House, 2000).

Smith, S., *Selected letters of Sydney Smith*, N.C. Smith (ed.) (Oxford: Oxford University Press, 1956).

Stewart, D., 'Preliminary Dissertation' or 'Dissertation first: exhibiting a general view of the progress of metaphysical, ethical and political philosophy since the revival of letters in Europe', M. Napier (ed.), *Encyclopaedia Britannica or Dictionary of the Arts, Sciences and General Literature* (21 vols, 7th edn, Edinburgh: A. and C. Black, 1842), I.

——, *Collected Works*, W. Hamilton (ed.) (11 vols, Edinburgh: Thomas Constable and Company; London: Hamilton, Adams and Company, 1854–60), II–IV.

Taylor, J., 'On the economy of the mines of Cornwall and Devon', *Transactions of the Geological Society*, Ser.I, II (1814), 309–328.

Townsend, J., *Dissertation on the Poor Laws by a Well-Wisher to Mankind* (London, 1786).

——, *A Journey Through Spain in the Years 1786 & 1787, With particular attention to the Agriculture, Manufactures, Commerce, Population, Taxes and Revenues of That Country and Remarks in passing Through A Part of France* (3 vols, London: C. Dilly, 1791), I–III.

——, *Character of Moses Established for Veracity as an Historian Recording Events from the Creation to the Deluge* (2 vols, Bath, 1813 and 1815), I–II.

Virgil, *The Georgics* translated and edited by L.P. Wilkinson (Harmondsworth: Penguin, 1982).

——, *The Eclogues* trans. G. Lee (Harmondsworth: Penguin, 1984).

Watt, G., 'Observations on Basalt & on the transition from vitreous to the stony texture', *Philosophical Transactions of the Royal Society of London* XCIV (1804) 279–314.

Wright, R. *An Apology for Dr Michael Servetus including an Account of his Life, Persecution, Writings and Opinions, being Designed to Eradicate Bigotry and Uncharitableness & to Promote Liberality of Sentiment among Christians* (Wisbech, 1806).

Printed secondary works

Allsobrook, D.I., *Schools for the Shires: The Reform of Middle-Class Education in Mid-Victorian England* (Manchester: Manchester University Press, 1988).

Armytage, W.H.G., 'Charles Watson-Wentworth, second Marquess of Rockingham, F.R.S (1730–1782): some aspects of his scientific interests', Notes and Records of the Royal Society of London, 12 (1956), 64–76.

Ashworth, W.J., 'John Herschel, George Airy, and the Roaming Eye of the State', *History of Science*, 26 (1998), 151–178.

——, ' "System of Terror": Samuel Bentham, accountability and dockyard reform during the Napoleonic Wars', *Social History*, 23 (1998), 63–79.

——, *Customs and Excise: Trade, Production and Consumption in England 1640–1845* (Oxford: Oxford University Press, 2003).

Aspinall, A., *Lord Brougham and the Whig party* (Manchester: University of Manchester, 1927).

——, 'The Coalition Ministries of 1827', *English Historical Review*, XLII (1927), 201–26, 533–539.

Aston, N., 'Glasse, Samuel (1734–1812)', *Oxford Dictionary of National Biography* (61 vols, Oxford: Oxford University Press, 2004), XXII.

Aubert, T., 'Alexander Aubert, F.R.S., Astronome, 1730–1805', *Notes and Records of the Royal Society of London*, 9 (1951), 79–95.

Ayling, S., *The Elder Pitt, Earl of Chatham* (London: Collins, 1976).

Ball T., Farr J., and Hanson, R.L. (eds), *Political Innovation and Conceptual Change* (Cambridge: Cambridge University Press, 1989).

Barnes, B. and Shapin, S., 'Science, nature and control: interpreting Mechanics' Institutes', *Social Studies of Science*, 7 (1977), 31–74.

Barrell, J., *The Dark Side of the Landscape: The Rural Poor in English Painting, 1730–1840* (Cambridge: Cambridge University Press, 1980).

Bayly, C.A., *Imperial Meridian: The British Empire and the World, 1780–1830* (London and New York: Longman, 1989).

——, *The birth of the Modern World, 1780–1914* (Malden, MA and Oxford: Blackwell, 2004).

Becker, H., *Outsiders: Studies in the Sociology of Deviance* (London: Free Press of Glencoe, 1963).

Belloni, L., 'Anatomica plastica: III. Die Florentiner Wachsplastiken', *Ciba Symposium*, 8 (1960), 129–132.

Bentley, M., *The Climax of Liberal Politics: British Liberalism in Theory and Practice, 1868–1918* (London: Edward Arnold, 1987).

——(ed.), *Public and Private Doctrine: Essays in British History Presented to Maurice Cowling* (Cambridge: Cambridge University Press, 1993).

Berman, M., *Social Change and Scientific Organisation: The Royal Institution, 1799–1844* (London: Heinemann Educational, 1978).

Bermingham, A., *Landscape and ideology: the English Rustic tradition, 1740–1860* (London: Thames and Hudson, 1987).

Bord, J., 'Patronage, the lansdowne Whigs and the problem of the liberal centre, 1827–8', *English Historical Review*, CXVII (2002), 78–93.

——, 'Whiggery, science and administration: Grenville and Lord Henry Petty in the ministry of all the talents, 1806–7', *Historical Research*, 76 (2003), 108–127.

Bowler, P., 'The Whig interpretation of Geology', *Biology and Philosophy*, 3 (1988), 99–103.

Brent, R., *Liberal Anglican Politics: Whiggery, Religion and Reform, 1830–41* (Oxford: Clarendon, 1987).

Brewer, J., *The Sinews of Power: War, Money and the English State, 1688–1783* (London: Unwin Hyman, 1989).

Brock, W.H., *Justus von Liebig: The Chemical Gatekeeper* (Cambridge: Cambridge University Press, 2002).

Brock, W.J. and Meadows, A.J., *The Lamp of Learning: Taylor and Francis and the Development of Science Publishing* (London: Taylor and Francis, 1998).

Browning, Reed, *The Duke of Newcastle* (New Haven CT, and London: Yale University Press, 1975).

——, *Political and Constitutional Ideas of the Court Whigs* (Baton Rouge LA, and London: Louisiana State University Press, 1982).

Burrow, J.W., *A Liberal Descent. Victorian Historians and the English Past* (Cambridge: Cambridge University Press, 1981).

——, *Whigs and Liberals: Continuity and Change in English Political Thought* (Oxford: Clarendon, 1988).

Butler, P. *Gladstone: Church, State and Tractarianism: A Study of his Religious Ideas and Attitudes, 1809–59* (Oxford: Clarendon, 1982).

Butterfield, H., *The Whig Interpretation of History* (London: G. Bell and Sons, 1931).
Bynum, W.F., Browne, E.J., and Porter, R. (eds), *Dictionary of the History of Science* (London and Basingstoke: Macmillan, 1981).
Cannadine, D., *Aspects of Aristocracy: Grandeur and Decline in Modern Britain* (London: Penguin, 1995).
Cannon, S.F., *Science in Culture: The Early Victorian Period* (Folkestone: Dawson, 1978).
Cantor, G.N., 'Henry Brougham and the Scottish methodological tradition', *Studies in the History and Philosophy of Science*, 2 (1971), 69–89.
——, 'The academy of Physics at Edinburgh, 1797–1800', *Social Studies of Science*, 5 (1975), 109–134.
Chalaby, J.K., 'Beyond the prison-house of language: discourse as a sociological concept', *British Journal of Sociology*, 47 (1996), 684–698.
Chitnis, A.C., *The Scottish Enlightenment and Early Victorian English Society* (London: Croom Helm, 1986).
Clapham, J.H., *An Economic History of Modern Britain* (3 vols, Cambridge: Cambridge University Press, 1926–38), I–II.
Clark, J.C.D, 'A general theory of party, opposition and government, 1688–1832', *Historical Journal*, 23 (1980), 295–325.
——*English Society 1660–1832: Religion, Ideology and Politics during the Ancien Regime* (2nd edn, Cambridge: Cambridge University Press, 2000).
Clark, J.F.M., 'History from the ground up: bugs, political economy and god in Kirby and Spence's *Introduction to Entomology*, 1825–1856', *Isis*, 97 (2006), 28–55.
Clive, J., *Scotch Reviewers; The Edinburgh Review, 1802–1815* (London: Faber and Faber, 1957).
Coley, N.G., Roberts, G.K., and Russell, C.A., *Chemists by Profession: The Origins and Rise of the Royal Institute of Chemistry* (Milton Keynes: Open University Press for the Royal Institute of Chemistry, 1977).
Collini, S., Whatmore, R., and Young, B. (eds), *History, Religion and Culture: British Intellectual History 1750–1950* (Cambridge: Cambridge University Press, 2000).
Collini, S., Winch, D., and Burrow, J., *That Noble Science of Politics: A Study in Nineteenth-Century Intellectual History* (Cambridge: Cambridge University Press, 1983).
Cookson, J.E., *The Friends of Peace: Anti-War Liberalism in England, 1793–1815* (Cambridge: Cambridge University Press, 1982).
Corsi, C., 'The Heritage of Dugald Stewart: Oxford philosophy and the method of political economy', *Nuncius*, 2 (1987), 89–144.
Crane, V.W., *The Club of Honest Whigs*: *The William and Mary Quarterly*, 23, (1966), 210–233.
Dalglish, J., *Eight Metaphysical Poets* (Oxford: Heinemann, 1961).
Daston, L., 'The ideal and reality of the republic of letters in the enlightenment', *Science in Context*, 4 (1991), 367–386.
Daston, L. and Otte, M., 'Introduction', *Science in Context*, 4 (1991), 223–232.
Dean, D.R., *James Hutton and the History of Geology* (Ithaca and London: Cornell University Press, 1992).
Desmond, A., *The Politics of Evolution: Morphology, Medicine, and Reform in Radical London* (Chicago and London: University of Chicago Press, 1989).
Dickinson, H.T., *Walpole and the Whig Supremacy* (London: English Universities Press, 1973).

——, *Liberty and Property: Political Ideology in Eighteenth Century Britain* (London: Weidenfeld and Nicolson, 1977).

Dinwiddy, J.R., 'Charles Fox as Historian', *Historical Journal*, 12 (1969), 23–34.

——, *Radicalism and Reform in Britain, 1780-1850* (London: Hambledon, 1992).

Drayton, R., *Imperial Science and Scientific Empire: Kew Gardens and the Uses of Nature, 1772–1903* (Ann Arbor, MI, 1993) [facsimile of PhD Thesis, Yale University, 1993].

——, *Nature's Government: Science, Imperial Britain, and the 'improvement' of the World* (New Haven and London: Yale University Press, 2000).

Eastwood, D. and Eastwood, D., 'Robert Southey and the intellectual origins of romantic conservatism', *English Historical Review*, CIV (1989), 308–331.

—— 'Robert Southey and the meanings of patriotism', *Journal of British Studies*, 31 (1992), 265–287.

Farkas, A., *State Learning and International Change* (Ann Arbor, MI: University of Michigan Press, 1998), pp. 72, 170.

Fisch, M. and Schaffer, S. (eds), *William Whewell, A Composite Portrait* (Oxford: Clarendon, 1991).

Fitzmaurice, E., *Life of William, Earl of Shelburne* (2 vols, London: Macmillan and Company, 1912), II.

Fontana, B., *Rethinking the Politics of Commercial Society: The Edinburgh Review, 1802–1832* (Cambridge: Cambridge University Press, 1985).

Forbes, D., *The Liberal Anglican Idea of History* (Cambridge: Cambridge University Press, 1952).

——, 'Scientific Whiggism: Adam Smith and John Millar', *Cambridge Historical Journal*, 7 (1954), 643–670.

——, *Hume's Philosophical Politics* (Cambridge: Cambridge University Press, 1975).

Ford, T.H., *Chancellor Brougham and his World: A Biography* (Chichester: Barry Rose Law, 2001).

Furet, F., *Interpreting the French Revolution* trans. E. Forster (Cambridge: Cambridge University Press, 1981).

Fruton, J.S., *Contrasts in Scientific Style: Research Groups in the Chemical and Biochemical Sciences* (Philadelphia: American Philosophical Society, 1990).

Galbraith, J.S., 'British war aims in World War I: a commentary on "Statesmanship" ', *Journal of Imperial and Commonwealth History*, 13 (1984), 25–45.

Gascoigne, J., *Science in the Service of Empire: Joseph Banks, The British State and the Uses of Science in the Age of Revolution* (Cambridge: Cambridge University Press, 1998).

——, 'Joseph Banks and his abiding legacy', *London Papers in Australian Studies*, 2 (2001), 1–12.

——, *Joseph Banks and the English Enlightenment: Useful Knowledge and Polite Culture* (Cambridge: Cambridge University Press, 2003).

Gash, N., *Reaction and Reconstruction in English Politics, 1832–1852* (Oxford: Clarendon Press, 1965).

——, *Aristocracy and People, Britain 1815–1865* (revised edition, London: Edward Arnold, 1983).

Geertz, C., *The Interpretation of Cultures: Selected Essays* (London: Fontana, 1993).

Geison, G.L. 'Scientific change, emerging specialties, and research schools', *History of Science*, 10 (1981), 20–40.

——, 'Research schools and new directions in the historiography of science', *Osiris*, 8 (1993), 227–238.
Gillispie, C.C., *Genesis and geology: A Study in the Relations of Scientific Thought, Natural Theology, and Social Opinion in Great Britain, 1790–1850* (Cambridge MA: Harvard University Press, 1951).
——, (ed.), *Dictionary of Scientific Biography* (16 vols, New York: Scribner, 1972), V Goddard, N., *Harvests of Change: The Royal Agricultural Society of England, 1838–1988* (London: Quiller Press, 1988).
Goldman, L., 'The Social Science Association, 1857–1886: a context for mid-Victorian liberalism', *English Historical Review*, CI (1986), 95–134.
——, 'The origins of British Social Science: political economy, natural science and statistics', *Historical Journal*, 26 (1983), 587–616.
Golinski, J., 'The theory of practice and the practice of theory: sociological approaches in the history of science', *Isis*, 81 (1990), 492–505.
——, *Science as Public Culture: Chemistry and Enlightenment in Britain, 1760–1820* (Cambridge: Cambridge University Press, 1992).
Goodwin, A., *The Friends of Liberty: The English Democratic Movement in the Age of the French Revolution* (Cambridge, MA, Harvard University Press, 1979).
Gould, S.J., 'More things in heaven and earth', in S. Rose (ed.), *The Richness of Life: The Essential Stephen Jay Gould* (London and New York, W.W. Norton and Company, 2007).
Guerrini, A., 'The Tory Newtonians: Gregory, Pitcairne and their circle', *Journal of British Studies*, 25 (1986), 288–311.
Hacking, I., *The Social Construction of What?* (Cambridge MA, and London: Harvard University Press, 1999).
Hall, A.R., 'On Whiggism', *History of Science*, 21 (1983), 45–59.
Harling, P., *The Waning of 'Old Corruption': The Politics of Economical Reform in Britain, 1779–1846* (Oxford: Clarendon Press, 1996).
Harrington, A., 'Interwar "German" psychobiology: between nationalism and the irrational', *Science in Context*, 4 (1991), 429–447.
Hays, J.N., 'Science and Brougham's Society', *Annals of Science*, 20 (1964), 227–241.
Hector, A. and Hooper, R., 'Darwin and the first ecological experiment', *Science*, 295, No. 5555, Issue of 25 January 2002, 639–640.
Hemingway, A., *Landscape Imagery and Urban Culture in Early Nineteenth-Century Britain* (Cambridge and New York: Cambridge University Press, 1992).
Hilton, B., *Corn, Cash, Commerce: The Economic Policies of the Tory governments, 1815–1830* (Oxford: Oxford University Press, 1977).
——, 'Whiggery, religion and social reform: the case of Lord Morpeth', *Historical Journal*, 37 (1994), 829–859.
——, *The Age of Atonement: The Influence of Evangelicalism on Social and Economic Thought, 1785–1865* (2nd edn Oxford: Clarendon Press, 1997).
——, *A Mad, Bad, and Dangerous People? England 1783–1846* (Oxford: Clarendon Press, 2006).
Himmelfarb, G., *The Idea of Poverty: England in the Early Industrial Age* (London: Faber, 1985).
Hoppit, J., 'Political arithmetic in eighteenth century England', *Economic History Review*, 49 (1996), 516–540.
——, 'Reforming Britain's Weights and Measures, 1660–1824', *English Historical Review*, CVIII (1998), 82–104.

Houghton, W.E. (ed.), *The Wellesley Index to Victorian Periodicals: Tables of Contents and Identification of Contributors with Bibliographies of their Articles and Stories* (5 vols, London and Toronto: University of Toronto Press, 1966–79), I.

Inkster, I. and Morrell, J. (eds), *Metropolis and Province: Science in British Culture, 1780–1850* (London: Hutchinson, 1983).

Jacob, M.C., *The Newtonians and the English Revolution, 1689–1720* (Hassocks: Harvester Press, 1976).

Jacyna, L.S., 'Immanence or Transcendence: theories of life and organisation in Britain, 1790–1835', *Isis*, 74 (1983), 311–329.

——, *Philosophic Whigs: Medicine, Science and Citizenship in Edinburgh, 1789–1848* (London and New York: Routledge, 1994).

——, and Stewart, L., *Practical Matter: Newton's Science in the Service of Industry and Empire, 1687–1851* (Cambridge MA, and London: Harvard University Press, 2004).

James, P., *'Population' Malthus: His Life and Times* (London: Routledge and Kegan Paul, 1979).

Jenkins, T.A., *Gladstone, Whiggery and the Liberal Party, 1874–1886* (Oxford: Clarendon, 1988).

Jones, P.M., 'Living the enlightenment and the French Revolution: James Watt, Matthew Boulton, and their sons', *Historical Journal*, 42 (1999), 157–182.

Jones, W.D., *Lord Derby and Victorian Conservatism* (Oxford: Basil Blackwell, 1956).

Jordanova, L.J. and Porter, R. (eds), *Images of the Earth: Essays in the History of the Environmental Sciences* (Chalfont St. Giles, British Society for the History of Science, 1979).

Jupp, P., *Lord Grenville: 1759–1834* (Oxford: Clarendon, 1985).

Kimizuka, N., 'Elder Statesmen and British party politics: Wellington, Lansdowne and the ministerial crises in the 1850s', *Parliamentary History*, 17 (1998), 355–372.

Knight, D., 'Agriculture and chemistry in Britain around 1800', *Annals of Science*, 33 (1976), 187–196.

Kramnick, I., *Bolingbroke and his Circle: The Politics of Nostalgia in the Age of Walpole* (Cambridge, MA: Harvard University Press, 1968).

Kriegel, A., 'Liberty and Whiggery in early nineteenth-century England', *Journal of Modern History*, 52 (1980), 253–278.

——, 'Eighteenth-century science and radical social theory: the case of Joseph Priestley's scientific liberalism', *Journal of British Studies*, 25 (1986), 1–30.

Langford, P., *A Polite and Commercial People: England, 1727–1783* (New York and Oxford: Oxford University Press, 1992).

Laudan, R., 'Ideas and organizations in British geology: a case study in institutional history', *Isis*, 68 (1977), 527–538.

——, *From Mineralogy to Geology: The Foundations of a Science, 1650–1830* (Chicago: University of Chicago Press, 1987).

Lee, S.M, 'A new language in Politicks; George Canning and the idea of opposition, 1801–1807' *History*, 83 (1998), 472–496.

Lemaine, G., Macleod, R., Mulkay, M., and Weingart, P. (eds), *Perspectives on the Emergence of Scientific Disciplines* (The Hague: Mouton, 1976).

Lilla, M., *G.B. Vico: The Making of an Anti-Modern* (Cambridge MA and London: Harvard University Press, 1994).

Mandler, P., *Aristocratic Government in the Age of Reform: Whigs and Liberals, 1830–5* (Oxford: Clarendon, 1990).
Mansfield, H., *Statesmanship and Party Government: A Study of Burke and Bolingbroke* (Chicago and London: University of Chicago Press, 1965).
Mayr, E., *The Growth of Biological Thought* (Cambridge MA: Belknap Press, 1982).
——, 'When is historiography Whiggish?', *Journal of the History of Ideas*, 51 (1990), 301–309.
McEvoy, J.G., 'Positivism, Whiggism, and the chemical revolution: a study in the historiography of chemistry, *History of Science*, 35 (1997), 1–33.
McKendrick, N. (ed.), *Historical Perspectives; Essays in Honour of J.H. Plumb* (London: Europa, 1974).
McNeil, M., *Under the Banner of Science: Erasmus Darwin and his Age* (Manchester: Manchester University Press, 1987).
Miller, D.P., 'The 'Hardwicke Circle': The Whig supremacy and its demise in the 18th- century royal society', *Notes and Records of the Royal Society of London*, 52 (1998), 73–91.
—— and Reill, P.H. (eds), *Visions of Empire: Voyages, Botany and Representations of Nature* (Cambridge: Cambridge University Press, 1996).
Miller, P.N. (ed.) *Joseph Priestley: Political Writings* (Cambridge: Cambridge University Press, 1993).
Mingay, G.E., *English Landed Society in the Eighteenth Century* (London: Routledge and Paul, 1963).
Mitchell, A., *The Whigs in Opposition, 1815–1830* (Oxford: Clarendon Press, 1967).
Mitchell, L.G., *C.J. Fox and the Disintegration of the Whig Party* (London: Oxford University Press, 1971).
——, *Holland House* (London: Duckworth, 1980).
——, 'Foxite politics and the great reform bill', *English Historical Review*, CVIII (1993), 338–364.
——, *Charles James Fox* (London: Penguin, 1997).
——, *Lord Melbourne, 1779–1848* (Oxford, Oxford University Press, 1997).
——, *The Whig World, 1760–1837* (London: Hambledon, 2005).
Mollon, J.D., 'The origins of the concept of interference', *Philosophical Transactions of the Royal Society of London*, 360 (2002), 807–819.
Money, J., 'Joseph Priestley in cultural context: Philosophic spectacle, popular belief and popular politics in eighteenth century Birmingham', *Enlightenment and Dissent*, 7 (1988), 57–81; 8 (1989), 69–89.
Moore, J.R. (ed), *History, Humanity and Evolution: Essays for John C. Greene*. (Cambridge: Cambridge University Press, 1989).
Morrell, J.B., 'Professors Robison and playfair and the *Theophobia Gallica*: natural philosophy, religion and politics in Edinburgh, 1789–1815', *Notes and Records of the Royal Society of London*, 26 (1971), 43–63.
——, 'The chemist breeders: the research schools of Liebig and Thomson', *Ambix*, 19 (1972), 1–46.
——, 'London Institutions and Lyell's Career: 1820–41', *British Journal for the History of Science*, 9 (1976), 132–146.
Morrell, J.B. and A. Thackray, *Gentlemen of Science: Early Years of the British Association for the Advancement of Science* (Oxford: Clarendon, 1981).

Morris, A.D., 'The Reverend Joseph Townsend MA MGS (1739–1816): physician & geologist – "Colossus of Rhodes" ', *Proceedings of the Royal Society of Medicine*, 62 (1969), 471–477.

Munday, P., 'Liebig's metamorphosis: from organic chemistry to the chemistry of agriculture', *Ambix*, 38 (1991), 135–154.

Namier, L., *The Structure of Politics at the Accession of George III* (2nd edn, London: Macmillan and Company, 1957).

New, C.W., *The Life of Henry Brougham to 1830* (Oxford: Clarendon Press, 1961).

Norman, E.R., *Anti-Catholicism in Victorian England* (London: Allen and Unwin, 1968).

Norris, J., *Shelburne and Reform* (London: Macmillan and Company, 1963).

Nutton, V. and Porter, R. (eds) *The History of Medical Education in Britain* (Amsterdam: Rodopi, 1995).

O'Connor, R., 'Mammoths and Maggots: Byron and the Geology of Cuvier', *Romanticism*, 5 (1999), 26–42.

Oldroyd, D.R., 'Historicism and the rise of historical geology, *History of Science*, 17 (1979), 191–213, 227–257.

Otte, M., 'Style as a historical category', *Science in Context*, 4 (1991), 233–264.

Owen, J.B., *The Rise of the Pelhams* (London: Methuen and Company, 1957).

Parker, R.A.C. 'Coke of Norfolk and the agrarian revolution', *Economic History Review*, 2nd series, 8 (1955–56), 156–166.

Parry, J., *The Rise and Fall of Liberal Government in Victorian Britain* (New Haven CT, and London: Yale University Press, 1993).

Pearce, C.D., 'Lord Brougham's neo-paganism', *Journal of the History of Ideas*, 55 (1994), 651–670.

Peltonen, M. (ed.), *The Cambridge Companion to Bacon* (Cambridge: Cambridge University Press, 1996).

Plumb, J.H., *Sir Robert Walpole: The King's Minister* (London: The Cresset Press, 1960).

Pocock, J.G.A., *Virtue, Commerce and History: Essays on Political Thought and History, Chiefly in the Eighteenth Century* (Cambridge: Cambridge University Press, 1985).

——, *The Machiavellian Moment: Florentine Political Thought and the Atlantic Republican Tradition* (2nd edn, Princeton NJ, and London: Princeton University Press, 2003).

Pompa, L. *Vico: A study of the 'New Science'* (2nd edn, Cambridge: Cambridge University Press, 1990).

Poole, R., *Time's Alteration: Calendar Reform in Early Modern England* (London: University College London Press, 1998).

Poovey, M., *A history of the Modern Fact; Problems of Knowledge in the Sciences of Wealth and Society* (Chicago and London: University of Chicago Press, 1998).

Porter, D. and Porter, R. (eds), *Doctors, Politics and Society: Historical Essays* (Amsterdam: Rodopi, 1993).

Porter, R., 'Charles Lyell and the principles of the history of Geology', *British Journal for the History of Science*, 9 (1976), 91–103.

——, *The Making of Geology: Earth Science in Britain, 1660–1815* (Cambridge: Cambridge University Press, 1977).

——, 'Gentleman and geology: the emergence of a scientific career, 1660–1920', *Historical Journal*, 21 (1978), 809–836.

—— and Brewer, J. (eds), *Consumption and the World of Goods* (London: Routledge, 1993).

Porter, T., *The Rise of Statistical Thinking: 1820–1900* (Princeton NJ, and Guildford: Princeton University Press, 1986).

——, *Trust in Numbers: The Pursuit of Objectivity in Science and Public Life* (Princeton NJ, and Chichester: Princeton University Press, 1995).

Prest, J., *Lord John Russell* (London: Macmillan, 1972).

Rapp, D., 'the left-wing Whigs: Whitbread, the mountain and reform, 1809–1815', *Journal of British Studies*, 21 (1982), 35–66.

Rashid, S. 'Dugald Stewart, "Baconian" methodology, and political economy', *Journal of the History of Ideas*, 46 (1985), 245–257.

Rebecchi, M., 'Paolo Andreani, un viaggiatore illuminato tra il settecento e l'ottocento', *Acme*, 54 (2001), 143–167.

Reingold, N., 'The peculiarities of the Americans or are there national styles in the sciences?', *Science in Context*, 4 (1991), 347–366.

Richards, R.J., 'Instinct and intelligence in British natural theology: some contributions to Darwin's theory of the evolution of behavior', *Journal of the History of Biology*, 14 (1981), 193–229.

——, *Darwin and the Emergence of Evolutionary Theories of Mind and Behaviour* (London and Chicago: University of Chicago Press, 1987).

Roberts, G.K., 'The Establishment of the Royal College of Chemistry: an investigation of the Social context of early-Victorian Chemistry', *Historical Studies in the Physical Sciences*, 7 (1976), 437–485.

Robertson, J.C., 'A Bacon-facing generation: Scottish philosophy in the early nineteenth century', *Journal of the History of Philosophy*, 14 (1976), 37–49.

Robbins, C., *The Eighteenth-Century Commonwealthsman* (Cambridge, MA: Harvard University Press, 1959).

Robinson, E., 'An English Jacobin: James Watt, junior, 1769–1848', *Cambridge Historical Journal*, 11 (1955), 349–355.

Rubinstein, W.D., *Elites and the Wealthy in Modern British Society* (Brighton: Harvester, 1987).

Rudwick, M.J.S., 'The Foundation of the Geological Society of London: its scheme for co-operative research and its struggle for independence', *British Journal for the History of Science*, 4 (1963), 325–355.

——, *The Meaning of Fossils: Episodes in the History of Palaeontology* (London: Macdonald and Company, 1972).

——, 'Historical Analogies in the geological work of Charles Lyell', *Janus*, 64 (1977), 89–107.

——, *The Great Devonian Controversy: The Shaping of Scientific Knowledge among Gentlemanly Specialists* (Chicago: University of Chicago Press, 1985).

Rupke, N., *The Great Chain of History: William Buckland and the English School of Geology (1814–1849)* (Oxford: Clarendon, 1983).

——, *Richard Owen, Victorian Naturalist* (New Haven CT, and London: Yale University Press, 1994).

Schwoerer, L.G., 'William, Lord Russell, the making of a Martyr', 1683–1983', *Journal of British Studies*, 24 (1985), 41–68.

Secord, J.A., *Controversy in Victorian Geology: The Cambrian-Silurian Dispute* (Princeton NJ: Princeton University Press, 1986).

——, *Victorian Sensation: The Extraordinary Publication, Reception and Secret Authorship of Vestiges of the Natural History of Creation* (Chicago: University of Chicago Press, 2000).

Shapin, S., *A Social History of Truth: Civility and Science in Seventeenth Century England* (Chicago and London: University of Chicago Press, 1994).

Shryock, R.H., 'The History of Quantification in Medical Science', *Isis*, 52 (1961), 215–237.

Skinner, Q., *Reason and Rhetoric in the Philosophy of Hobbes* (Cambridge: Cambridge University Press, 1996).

——, *Liberty before Liberalism* (Cambridge: Cambridge University Press, 1998).

Smith, C. and Wise, M.N., 'Work and waste: political economy and natural philosophy in nineteenth century Britain (I)', *History of Science*, 27 (1989), 263–301, 391–449; 28 (1990), 221–261.

Smith, E.A., *Whig Principles and Party Politics: Earl Fitzwilliam and the Whig party, 1748–1833* (Manchester: Manchester University Press, 1975).

Solkin, D.H., *Richard Wilson: The Landscape of Reaction* (London: Tate Gallery, 1982).

Stauffer, R.C., *Charles Darwin's Natural Selection* (London: Cambridge University Press, 1975).

Stevenson, J., 'William Cobbett: patriot or Briton?', *Transactions of the Royal Historical Society*, 6th Series, 6 (1996), 123–136.

Stewart, L., *The Rise of Public Science: Rhetoric, technology and Natural Philosophy in Newtonian Britain, 1660–1750* (Cambridge: Cambridge University Press, 1992).

Stewart, R., *Henry Brougham, 1778–1868: his Public Career* (London: Bodley Head, 1985).

Stone, L., *An Imperial State at War: Britain from 1689 to 1815* (London: Routledge, 1994).

Teich, M. and Young, R. (eds), *Changing Perspectives in the History of Science: Essays in Honour of Joseph Needham* (London: Heinemann Educational, 1973).

Thomas, W., *The Philosophic Radicals: Nine Studies in Theory and Practice, 1817–1841* (Oxford: Clarendon Press, 1979).

Thompson, F.M.L., *English Landed Society in the Nineteenth Century* (London: Routledge and Kegan Paul; Toronto: University of Toronto Press, 1963).

Thorne, R.G., *History of Parliament: The Commons, 1790–1820* (5 vols, London: Secker and Warburg for the History of Parliament Trust, 1986), I–V.

Torrance Kirby, W.J., *The Theology of Richard Hooker in the Context of the Magisterial Reformation* (Princeton NJ, Princeton Seminary Press, 2000).

Torrens, H.S., *The Practice of British Geology, 1750–1850* (Aldershot: Ashgate Variorum, 2002).

Tully, J. (ed.), *Meaning and Context: Quentin Skinner and his Critics* (Princeton NJ: Princeton University Press, 1988).

Urquhart, R., 'The trade wind, the statesman and the system of commerce: Sir James Steuart's vision of political economy', *European Journal of the History of Economic Thought*, 3 (1996), 379–410.

Viroli, M., *Machiavelli* (Oxford and New York: Oxford University Press, 1998).

Wasson, E.A., 'The coalitions of 1827 and the crisis of Whig leadership', *Historical Journal*, 20 (1977), 587–606.

——, 'The Third Earl Spencer and agriculture, 1818–1845', *Agricultural History Review*, 26 (1978), 89–99.

——, *Whig Renaissance, Lord Althorp and the Whig Party, 1782–1845* (New York and London: Garland, 1987).

Waters, C.K., 'The arguments in the *Origin of Species*', J. Hodge and G. Radick (eds), *Cambridge Companion to Darwin* (Cambridge, 2003), pp. 119–127.

Wessely, A., 'Transposing "Style" from the history of art to the history of science', *Science in Context*, 4 (1991), 265–278.

Whalley, G. 'Coleridge and Vico', G. Tagliacozzo and H.V. White (eds), *Giambattista Vico: an International Symposium* (Baltimore: Johns Hopkins Press, 1969).

Whittingham-Jones, B., *Liverpool's Political Clubs, 1812–30* (Liverpool, 1959). Article also published in *Transactions of the Historic Society of Lancashire and Cheshire*, 111(1959) 117–138.

Wilkinson, L.P., *The Georgics of Virgil* (London: Cambridge University Press, 1969).

Winch, D., *Riches and Poverty: An Intellectual History of Political Economy in Britain, 1750–1834* (Cambridge: Cambridge University Press, 1996).

Wohlgemuth, M., 'Evolutionary approaches to politics', *Kyklos*, 55 (2002), 2, 223–246.

Woodward, H.B., *The History of the Geological Society of London* (London: Longmans, Green and Company, 1907).

Wyatt, J., 'George Bellas Greenough: a romantic geologist', *Archives of Natural History*, 22 (1995), 61–71.

——, *Wordsworth and the Geologists* (Cambridge: Cambridge University Press, 1995).

Yeo, R., 'An idol of the market-place: Baconianism in 19th century Britain', *History of Science*, 23 (1985), 251–298.

—— *Defining Science: William Whewell, Natural Knowledge and Public Debate in early Victorian Britain* (Cambridge: Cambridge University Press, 1993).

Unpublished dissertations

Roberts, G.K., 'The Royal College of Chemistry (1845–53): a social history of chemistry in early-Victorian England', MS. [microfilm] PhD. (The Johns Hopkins University, 1973).

Yule, J.D., 'The impact of science on British Religious thought in the second quarter of the nineteenth century', MS. PhD. (University of Cambridge, 1976).

Index

The letter 'n' denotes note numbers